Schriftenreihe aus dem Institut für Strömu

Herausgeber
J. Fröhlich, S. Odenbach, K. Vogeler

Institut für Strömungsmechanik
Technische Universität Dresden
D-01062 Dresden

Band 13

Claudio Santarelli

Direct Numerical Simulations of bubbles in turbulent flows with heat transfer

TUDpress

2015

Die vorliegende Arbeit wurde am 21. Januar 2015 an der Fakultät Maschinenwesen der Technischen Universität Dresden als Dissertation eingereicht und am 10. Juni 2015 erfolgreich verteidigt.

This work was submitted as a PhD thesis to the Faculty of Mechanical Science and Engineering of TU Dresden on 21 January 2015 and successfully defended on 10 June 2015.

Gutachter | Reviewers
Prof. Dr.-Ing. habil. Jochen Fröhlich, TU Dresden
Prof. Dr. ir. J.A.M. (Hans) Kuipers, TU Eindhoven

Bibliografische Information der Deutschen Nationalbibliothek
Die Deutsche Nationalbibliothek verzeichnet diese Publikation in der Deutschen Nationalbibliografie; detaillierte bibliografische Daten sind im Internet über http://dnb.d-nb.de abrufbar.

Bibliographic information published by the Deutsche Nationalbibliothek
The Deutsche Nationalbibliothek lists this publication in the Deutsche Nationalbibliografie; detailed bibliographic data are available in the Internet at http://dnb.d-nb.de.

ISBN 978-3-95908-017-0

© 2015 TUDpress
Verlag der Wissenschaften GmbH
Bergstr. 70 | D-01069 Dresden
Tel.: 0351/47 96 97 20 | Fax: 0351/47 96 08 19
http://www.tudpress.de

Technische Universität Dresden
Fakulät Maschinenwesen

Direct Numerical Simulations of bubbles in turbulent flows with heat transfer

Dissertation

zur Erlangung des akademischen Grades

Doktor-Ingenieur (Dr.-Ing.)

von

Claudio Santarelli

geboren am 24. August 1983

in Rom, Italien

Gutachter   Prof. Dr.-Ing. habil. Jochen Fröhlich
                             Prof. Dr. ir. J.A.M. (Hans) Kuipers

Tag der Einreichung: 21.01.2015
Tag der Verteidigung: 10.06.2015

Single-phase flow is so 20th century.

Jean Roussel, French Engineer-to-be

Acknowledgments

It is not easy to collect my thoughts concerning this great experience. So I just start from the beginning: This work would have not been possible without Professor Jochen Fröhlich. He knows as well as I do that it has not always been easy, but it has always been challenging. I am also very grateful to Professor Hans Kuipers for accepting to review this work and for the interesting discussion we had afterwards.

How not to mention my colleagues at the PSM? I have been very lucky to shear some years with them, both from a professional and from a personal point of view. My years in Dresden would have not been so valuable to me without the discussions we had and their support I could always rely on. Also the work with my students has been a great source of inspiration as well as a personal pleasure. I hope they could benefit from our work together as much as I did. The same holds for my friends, the ones who shared my everyday life and the ones "abroad": It helped me a lot knowing I could always count on them, in the joyful as well as in the difficult times. Thank you guys for all the wonderful experiences we shared.

Last but not least I want to thank my family. Even if we have been *geographically* distant, I always felt their presence and encouragement. And most of all, I know that this is never going to change. This is why I dedicate them this work and the result achieved.

Abstract

The present thesis addresses the dynamics of disperse air bubbles in turbulent channel flows under different conditions. Direct Numerical Simulations on a Cartesian grid are performed where the bubbles are introduced in the Navier-Stokes equation by means of an Immersed Boundary Method. This analysis first focuses on the flow of a swarm of small bubbles where the total void fraction is 2.14%. Instantaneous flow visualizations as well as statistical quantities for both the carrier and the disperse phase are reported and compared with the ones of an unladen channel flow.

Additional simulations of the same configuration allow addressing the influence of different parameters like the total void fraction, the bubble size, the bidispersity of the swarm and the direction of the flow direction. In this manner, a parameter analysis is performed where several flow features are investigated, such as bubble-induced fluid structures, turbulence quantities and bubble interaction, to mention but a few.

The role of the bubbles on the modification of the turbulence of the liquid phase is investigated by means of a budget analysis arising from the conservation equation of the turbulent kinetic energy for bubbly flows. Furthermore, the influence of the bubbles on the instantaneous local modification of the liquid turbulence is addressed by means of dedicated flow visualizations that are here presented for the first time.

Successively, to address the heat transfer in the two-phase flow, the original numerical code is improved to account for the temperature field of the mixture. Special attention is given to the boundary condition at the phase boundary of the bubbles and, to this end, two approaches are implemented and validated. This allows the investigation of the heat transfer between the two-phase flow and the walls in a channel flow configuration and of the complex phenomena involved.

The present work, hence, provides quantitative statistical data that can be employed for validation as well as for development and improvements of turbulence models in the framework of bubbly flows.

Zusammenfassung

Die vorliegende Dissertation beschäftigt sich mit der Dynamik disperser Luftblasen in turbulenten Kanalströmungen unter verschiedenen Bedingungen. Es werden Direkte Numerische Simulationen auf einem kartesischen Gitter durchgeführt, wo die dispersen Blasen mittels einer Immersed Boundary Methode in der Lösung der Navier-Stokes Gleichungen berücksichtigt werden. Diese Untersuchung beginnt mit der Analyse der Strömung eines Schwarmes kleiner Blasen, wobei der Gasgehalt 2.14% beträgt. Es werden sowohl Strömungsvisualisierungen als auch statistische Größen beider Phasen präsentiert und mit denen einer unbeladenen Strömung verglichen.

Der Einfluss verschiedenen Parameter wie der Gasgehalt, die Blasengröße, die Dispersität des Schwarmes und die Richtung der mittleren Geschwindigkeit des Fluids wird anhand zusätzlicher Simulationen untersucht. Dabei wird eine Parameterstudie durchgeführt, welche die gezielte Analyse von Strömungseigentschaften, wie z.B. blaseninduzierte Fluidstrukturen, Turbulenzgrößen und Interaktion der Blasen etc., ermöglicht.

Die Rolle der Blasen auf die Modifizierung der Turbulenz der flüssigen Phase wird anhand der Transportgleichung der turbulenten kinetischen Energie in Blasenströmungen untersucht. Dem lokalen, instantanen Einfluss der Blasen auf die Turbulenz werden Strömungsvisualisierungen der dazugehörige Bilanzterme gewidmet, die in dieser Art zum ersten Mal präsentiert werden.

Die ursprüngliche numerische Methode wird erweitert, um das Temperaturfeld der Blasenströmung und den daraus folgenden Wärmeaustauch zwischen dem Zweiphasengemisch und den festen Wänden des Kanals zu untersuchen. Besondere Aufmerksamkeit wird der thermischen Randbedingung auf der Phasengrenze der Blasen gegeben und zwei verschiedene Modelle werden implementiert und validiert. Dadurch ist eine Analyse der komplexen Phänomene möglich, die in solchen Strömungen eine tragende Rolle spielen.

Schließlich liefert die vorhandene Arbeit statistische und quantitative Daten, die der Verbesserung und Entwicklung numerischer Modelle im Rahmen von Blasenströmungen dienen können.

Riassunto

Questa tesi é dedicata allo studio della dinamica di bolle d'aria in un flusso turbolento in un canale piano verticale. L'analisi é condotta mediante Simulazioni Numeriche Dirette dove le bolle sono introdotte nella soluzione dell'equazioni di Navier-Stokes mediante un metodo di Immersed Boundary. Questo studio inizia con l'analisi del flusso di uno sciame di bolle piccole, dove la frazione di gas é pari al 2.14%. Sono presentate sia immagini istantanee del flusso sia quantità statistiche per entrambi le fasi e tali grandezze vengono confrontante con le rispettive grandezze di un flusso monofase.

Ulteriori simulazioni della stessa configurazione permettono di analizzare singolarmente il ruolo di diversi parametri come, ad esempio, la frazione di gas, la grandezza delle bolle, la bidispersione dello sciame e la direzione della velocità del fluido. In questo modo é possibile un'analisi comparativa di alcune grandezze di interesse fluidodinamico, come ad esempio la presenza strutture turbolente indotte dalle bolle, le statistiche della turbolenza e l'interazione tra le bolle.

Il ruolo delle bolle nella variazione delle turbolenza delle fase liquida é analizzato mediante l'analisi del bilancio dei termini dell'equazione di trasporto dell'energia cinetica turbolenta in flussi bifase. L'impatto locale ed istantaneo prodotto dalle bolle é analizzato mediante immagini, presentate per la prima volta, dei suddetti termini dell'equazione di bilancio.

Il codice di calcolo originale é ampliato per rendere possibile l'analisi dello scambio termico tra la miscela bifase e le pareti verticali nel canale. Particolare attenzione é dedicata alla corretta imposizione delle condizioni al contorno per la temperatura sulla superficie delle bolle e due diversi modelli numerici sono implementati e validati. In questo modo é resa possibile l'analisi dello scambio termico e dei fenomeni fluidodinamici coinvolti.

In conclusione, nella presente tesi vengono fornite grandezze statistiche che possono essere impiegate per il perfezionamento ed eventualmente per lo sviluppo di nuovi modelli per la simulazione di flussi bifase turbolenti.

Contents

Nomenclature

Roman Symbols

a	thermal diffusivity
C_c	constant in the collision model
d_p	bubble diameter
f	liquid volumetric fraction
$\mathbf{f}$	volume force with components f_x, f_y, f_z
H	distance between channel walls, $L_y = H$
I_w	width of the wall-normal bin for bubble statistics
K	turbulent kinetic energy
$\widetilde{K}$	local instant turbulent kinetic energy
K_Ω	heat-flux at the phase boundary
L_x, L_y, L_z	Cartesian dimensions in x-, y- and z-direction, respectively
l_τ	viscous length scale
N_f	number of flow fields for statistical analysis
N_p	total number of bubbles
N_w	number of wall-normal bins for bubble statistics
p	fluid pressure
q_w	heat flux at the wall
q_0	reference heat flux
Re_b	bulk Reynolds number
Re_p	bubble Reynolds number in the channel
$Re_{p,c}$	Reynolds number of fixed objects
$Re_{p,\infty}$	Reynolds number of rising bubbles
r_p	bubble radius
S	local instant interfacial area concentration
t	time
T	temperature
T_b	bulk time unit
T_v	magnitude of the turbophoresis effect
$\mathbf{u}$	velocity vector with Cartesian components u, v, w
U_b	bulk velocity
u_c	inflow velocity

$\mathbf{u}_p$	velocity vector with bubble velocity u_p, v_p, w_p
$\mathbf{u}_r$	relative bubble velocity, with components u_r, v_r, w_r
U_r	weighted relative bubble velocity
u_τ	friction velocity
$u_{p,\infty}$	bubble terminal rise velocity in quiescent fluid
V_p	bubble volume
$\mathbf{x}$	coordinate vector with components x, y, z
$\mathbf{x}_p$	coordinate vector of bubble position with components x_p, y_p, z_p
y_w	distance from wall, $y_w = \min\{y, H - y\}$

Greek Symbols

δ	distance between phase boundary and help points
Δ	step size of the simulation mesh
ΔE	quantification of hindrance effect
ϕ_{tot}	total void fraction
Φ	liquid indicator function
ρ	density of the fluid
ρ_p	density of the bubbles
ν	kinematic viscosity of the fluid
τ_w	wall shear stress
$\chi_{[a,b]}(x)$	indicator function, $\chi = 1$ if $x \in [a,b]$, $\chi = 0$ else
Ω	bubble surface
π_ρ	density ratio, ρ_p/ρ

Subscripts

2	two-dimensional
3	three-dimensional
c	collision-related quantity
b	bulk flow-related quantity
L	interfacial quantity
p	bubble-related quantity
r	relative velocity-related quantity
ref	reference quantity
w	wall-related quantities

Other symbols

$\langle\ldots\rangle_a$	averaging operator over direction a
$(\ldots)'$	fluctuation
$\overline{(\ldots)}$	single-phase averaged quantity
$\overline{\overline{(\ldots)}}$	phase-weighted averaged quantity
$\widetilde{(\ldots)}$	local instant quantity

Abbreviations

BC	boundary condition
DNS	Direct Numerical Simulation
IBM	Immersed Boundary Method
LES	Large Eddy Simulation
PB	phase boundary
PRIME	Phase Resolving sIMulation Environment
RANSE	Reynolds-averaged Navier-Stokes Equations
RDB	randomly distributed bubbles
RHS	right-hand side
RMS	root mean square

1 Introduction

1.1 Physics of bubbly flows

Two-phase flows and in particular bubbly flows are an essential part of many industrial applications, occurring in nuclear power generation, chemical industry, in food-processing industry and many other installations. As soon as a flow contains a second phase, complex phenomena regarding the interaction of the two phases take place and in general need to be accounted for. One of the most fascinating aspects is the mutual interaction between the bubbles and the surrounding fluid turbulence in a configuration as simple as a vertical bubble-laden pipe flow: The bubbles modify the turbulence and thus the behavior of the fluid which, in turn, influences the behavior of the bubbles themselves, resulting in a phenomenon of substantial complexity still not fully understood. Due to its relevance, this topic has been investigated repeatedly over the last decades and several approaches have been employed to address such phenomena (Clift et al. (2005), Crowe (2005), Michaelides (2006)). In view of the large body of work, the following literature review addresses only papers very closely related to the present study.

The first studies of bubbly flows were experimental investigations and among the fundamental contributions the work of Serizawa et al. (1975a,b) has to be mentioned. These authors investigated upward-directed bubbly flows in a vertical pipe and focused on the disperse regime and the slug flow regime. For the first one, they observed peaks of the radial void fraction distribution at the walls. Later on, Wang et al. (1987) extended these studies also to downward-directed bubbly, for which a peak of the void fraction in the center region was observed. In their experiments, the presence of bubbles could both enhance or diminish the liquid turbulence, depending on the flow rate and the global void fraction. Since then, many more experimental studies have been conducted for the investigation of different aspects of this type of flow. Takagi et al. (2008), for example, performed experiments of upward bubbly flow in a plane channel geometry and studied the influence of surfactants on the dynamics of the bubbles. They observed that half-contaminated bubbles experience a lift force, induced by the mean flow gradient, that pushes them toward the walls where they form horizontal clusters, while fully-contaminated bubbles do not experience the same lift force. Martinez Mercado et al. (2010) devoted their study to the so-called pseudo-turbulence, that is when bubbles rise in an otherwise quiescent fluid and turbulence is generated only by the bubbles. In their work bubble-bubble interaction was investigated and the influence of the bubbles on the fluid energy spectra was addressed. A slope of -3.2 of the double-logarithmic energy spectrum of the velocity fluctuation was observed which is in close agreement with (Mendez-Diaz et al., 2013) for a similar configuration. In a previous work Mendez-Diaz et al. (2012) collected several experimental results to develop a criterion for the void fraction distribution as a function of the bubble Reynolds number and of the Weber number.

At the same time, numerical investigations were carried out by other researchers in a strong relationship with experiments. The first computations of bubbly flows were based on models for the description of the mean quantities of the flow, in what now is commonly referred to as two-fluid models. Among the first approaches, the one proposed by Sato et al. (1981) should be mentioned. This model is based on the assumption that the fluid turbulence can be decomposed in two contributions, one related to the fluid itself induced by a mean shear and one related to the presence of bubbles. This assumption is still one of the most employed for modeling bubbly flows (Lopez de Bertodano et al., 1994; Troshko and Hassan, 2001; Krepper et al., 2005). Such models are based on correlations that were developed from experimental results, either derived from the analysis of the dynamics of a single bubble or from the investigation of bubble swarms. During the last years, increasing computer resources have allowed performing more accurate simulations, where the bubble dynamics and the coupling between the two phases do not rely on models but are resolved by the numerical method. These computational approaches, mainly referred to as two-phase Direct Numerical Simulations (DNS), have proven to be a trustworthy tool to gain deep understanding of the complex phenomena involved. Furthermore, DNS of bubbly flows can be used to deduce closure relations for two-fluid models as described, for example, by Deen et al. (2004).

To simulate a large number of bubbles, mostly two configurations have been investigated in the literature: A triply periodic domain and channel flow. The first configuration is used to investigate the behavior of bubble swarms in an unbounded infinite domain. This design was employed, among others, by Bunner and Tryggvason (2002a,b) for nearly spherical bubbles and later by the same authors for deformable bubbles, focusing on bubble-bubble interaction and on the spectrum of the velocity fluctuations (Bunner and Tryggvason, 2003). Their analysis was later extended to bubbles at higher Reynolds number in (Esmaeeli and Tryggvason, 2005). Roghair et al. (2011) performed simulations of bubble swarms in the periodic domain to investigate the drag of the bubbles in monodisperse swarms and to compare it to the drag experienced by a single bubble. They derived a correlation for the bubble drag coefficient as a function of the void fraction and in a later work expanded the analysis to bidisperse swarms (Roghair et al., 2013). Although valuable information can be obtained from these simulations, results are influenced by the periodicity constraint and by the dimension of the domain, which is not able to capture the large-scale flow features that can have a strong influence on the small-scales behavior. An example of such an influence is given in (Roghair et al., 2011), where the numerical results regarding the bubble pair alignment were compared with the experimental works of Martinez Mercado et al. (2010). Poor agreement was observed due to the "lack of large-scale flow circulations due to the limited size of the computational domain and absence of walls in the domain" (Roghair et al., 2011).

The second geometrical configuration employed for the analysis of bubble swarms is channel flow, where the two-phase flow is confined between two vertical walls and is periodic in the streamwise and spanwise direction. The advantages of this setup are a larger resemblance to real industrial situations and, as in the case of the periodic domain, the possibility to collect statistics over time and over the two periodic directions. Additionally, the presence of the walls increases the complexity of the flow, due to the presence of a mean shear flow of the fluid velocity and due to the interaction between bubbles and wall turbulence. Examples of studies regarding the interaction between bubbles and walls are the work of Takemura and Magnaudet (2003), who performed experiments for bubbles rising along a vertical wall, and the work of Tran-Cong et al. (2008), who studied the influence of a turbulent boundary layer on the dynamics of small bubbles. Regarding the simulation of bubble swarms in vertical channels, fundamental contributions were provided by Tryggvason and his research team. Lu et al. (2006) investigated bubble swarms in laminar channel flow, both for upward and downward flow configurations. For the upward case, they found that bubbles tend to rise in the wall region and that in the downward case bubbles tend to rise in the center region. Both features are in good agreement with experimental observations (Serizawa et al., 1975a; Wang et al., 1987). Lu and Tryggvason (2008) performed simulations of nearly spherical and of deformable bubbles in a turbulent vertical channel flow and observed a different behavior of the two bubble classes. While nearly spherical bubbles tend toward the walls as in the laminar case, the deformable ones prevalently rise in the channel center. This was also observed in experiments conducted by Tomiyama et al. (2002) investigating the rise of single bubbles in flow with constant shear rate. The different behavior was hypothesized to be due to the effect of the slanted wake generated by the deformable bubble, which causes the change of sign of the lift force. A more detailed investigation was later provided by Adoua et al. (2009) by means of DNS of the flow around an oblate ellipsoid with a body-fitted mesh. Two different processes for the genesis of the lift force were observed, one related to the shear flow and the other related to the vorticity generation at the phase boundary. The latter effect was found dominant for ellipsoidal bodies determining, hence, the sign of the lift force.

The influence of deformability on the dynamics of bubble swarms was further addressed by Dabiri et al. (2013) who confirmed the aforementioned behavior of bubbles with different deformations. Lu and Tryggvason (2013) extended their previous studies and simulated a bubble swarm where the background fluid turbulence was increased with respect to the earlier study (Lu and Tryggvason, 2008). Tanaka (2011) performed similar computations of a swarm of nearly spherical bubbles, focusing on the heat transfer mechanism between the two-phase mixture and the walls. In the same configuration, Bolotnov et al. (2011) investigated a swarm of nearly spherical bubbles and observed that bubbles tend to rise in the wall region and to increase the production of turbulence. Yamamoto and Kunugi (2011) analyzed the behavior of four bubbles at somewhat larger bubble Reynolds number, around 120. They noted that two of the bubbles stayed mainly in the center region and that one bubble rose in the near-wall region, possibly captured by the low-pressure zone of high-speed streaks, and no information was given on the fourth bubble.

Bubbles have proven to modify not only small-scale flow features but also the large scales. Such an influence was described in the pioneering work of Lance and Bataille (1991) for upward bubbly flow in grid-generated turbulence and afterward by Panidis and Papailiou (2000). A similar feature was observed by Uhlmann (2008) and Garcia-Villalba et al. (2012) for the sedimentation of heavy particles in channel flow, where the presence of elongated structures is due to an intrinsic instability of the flow triggered by the presence of the particles. Many of the DNS of bubbly flow mentioned here were performed in small channel geometries due to limited computer resources and therefore large-scale flow features could not be addressed. As a result, a clear separation of flow scales was not possible: Small scales influence the large ones and vice versa. Several authors state that neglecting the large-scale flow features can have a substantial influence on the results (Tanaka, 2011; Roghair et al., 2011; Dabiri et al., 2013). One of the goals of the present work, therefore, is to address the large-scale flow features generated by the bubbles and to investigate the influence of these on the bubble dynamics.

This overview highlights the great research effort that has been devoted in the last decades to this wide topic and the importance of this research field that continuously motivates new studies. Despite the awareness of the complexity of such phenomena, the present work is intended to contribute to shed some light on some of the aforementioned issues focusing on selected features of bubbly flows.

1.2   Thesis outline

The present thesis is organized as follows. Chapter 2 is devoted to the numerical approaches employed for the simulation of single-phase and two-phase flows and, afterward, the Immersed Boundary Method employed in this work for the simulation of bubbly flow is presented. In Chap. 3 the configuration investigated and the whole set of simulations performed is introduced. A validation of the numerical method employed is provided and results are presented for the simulation of a dense swarm of small bubbles, labeled *SmMany*, which will be referred to as the reference case throughout the present work. Chapter 4 addresses the influence of the total void fraction of the flow and, to this end, a simulation labeled *SmFew* is introduced where a dilute swarm of small bubbles is considered. In Chap. 5 two simulations are presented to address the influence of the bubble size and a validation of the numerical method is presented for large bubbles. The simulation labeled *LaMany*, where a dense swarm of large bubble is simulated, allows investigating the influence of the bubble size and, eventually, in the *BiDisp* case a bidisperse swarm of small and large bubbles is investigated to analyze the influence of the bidispersity. In Chap. 6 a simulation labeled *SmManyDo* in a downward configuration is addressed, where the direction of the fluid velocity is opposed with respect to the reference case *SmMany*. Chapter 7 presents a detailed analysis of the modification of the turbulence of the liquid phase induced by the bubbles by means of the budget analysis of the transport equation of the liquid turbulent kinetic energy and by means of dedicated flow visualizations. In Chap. 8 the development and implementation of a *thermal* Immersed Boundary Method is presented and employed for the investigation of the heat transfer between the walls and the two-phase mixture in the channel flow. The thesis ends with Chap. 9 where the main achieved results are collected and an outlook on possible future research is provided.

2 Computational method

2.1 Simulation of single-phase flow

Direct Numerical Simulation

In a Direct Numerical Simulation (DNS) all flow scales, both spatial and temporal, are resolved and the mesh step size and the time step must be fine enough to capture all flow features. This strategy allows avoiding any models for the smallest scales, often referred to as the turbulent scales. This means that no additional term has to be considered in the Navier-Stokes Equations (NSE). For incompressible fluids, as the ones considered in the present work, the NSE read

$$\frac{\partial u_i}{\partial x_i} = 0 \tag{2.1}$$

$$\frac{\partial u_i}{\partial t} + \frac{\partial (u_i\, u_j)}{\partial x_j} = -\frac{1}{\rho}\frac{\partial p}{\partial x_i} + \frac{1}{\rho}\frac{\partial \tau_{ij}}{\partial x_j} + f_i \tag{2.2}$$

where u_i is the fluid velocity component in the i direction, p is the fluid pressure with the hydrostatic part being subtracted, f_i collects all volume forces and τ_{ij} is the stress tensor defined as

$$\tau_{ij} = \mu \left(\frac{\partial u_j}{\partial x_i} + \frac{\partial u_i}{\partial x_j} \right) \tag{2.3}$$

and where μ is the fluid dyanamic viscosity where $\nu = \mu/\rho$ is the fluid kinematic viscosity. Repeated indexes imply summation according to the Einstein notation.

For turbulent flows the size of the smallest eddies can be quantified by the Kolmogorov length scale η defines as

$$\eta = \left(\frac{\nu^3}{\epsilon} \right)^{1/4} \tag{2.4}$$

and the temporal scale associated to such scales is the Kolmogorov time scale, defined as

$$\tau_\eta = \left(\frac{\nu}{\epsilon} \right)^{1/2} , \tag{2.5}$$

where ϵ is the dissipation of the turbulent kinetic energy, that will be introduced later. On the other hand, the largest scales of the flow may be represented by a characteristic velocity U and a characteristic or integral length L so that a Reynolds number can be defined as

$$Re = \frac{U\, L}{\nu} \tag{2.6}$$

to provide a qualitative classification of the flow. Furthermore, it can be shown with dimensional analysis that the ratio between the smallest and the integral scale, η/L, behaves as

$Re^{-3/4}$ (Pope, 2000). Considering a three dimensional numerical grid, which is essential to capture the three-dimensionality of turbulence, the total number of grid points needed for a DNS then scales with $Re^{9/4}$ and this easily gives an impression on how costly DNS are from a computational point of view. To overcome this problem and to perform simulations of flows at high Reynolds numbers, two main strategies can be mentioned: Large Eddy Simulations and Reynolds Averaged Navier-Stokes Equations, both addressed in the following sections.

Large Eddy Simulation

The fundamental idea of Large Eddy Simulations (LES) is to simulate the large eddies and to to model the small ones. This concept is motivated by the observation that the small scales present a common behavior for all types of flows, while the large flow scales strongly depend on the flow configuration (e.g. geometry, boundary conditions). Formally, the velocity can be divided into a resolved or filtered term $\widehat{u}_i$ and an unresolved or sub-grid component $u_i^{''}$ as

$$u_i = \widehat{u}_i + u_i^{''} \tag{2.7}$$

where $\widehat{u}_i$ is defined by means of a filtering operation as

$$\widehat{u}_i(\mathbf{x}) = \int_{\mathbb{R}^1} G(\mathbf{x}, \mathbf{y}, \Delta^{''}(\mathbf{x}))\, u_i(\mathbf{y})\, d\mathbf{y} \,. \tag{2.8}$$

Here the dependence of $\widehat{u}_i$ on the position $\mathbf{x}$ is explicitly introduced in the notation and volume forces are neglected for clarity. The quantity G is the filter kernel and $\Delta^{''}$ the width of the filter. Substituting (2.7) into (2.1) and (2.2) yields

$$\frac{\partial \widehat{u}_i}{\partial x_i} = 0 \tag{2.9}$$

and

$$\frac{\partial \widehat{u}_i}{\partial t} + \frac{\partial (\widehat{u}_i\, \widehat{u}_j)}{\partial x_j} = -\frac{1}{\rho}\frac{\partial \widehat{p}}{\partial x_i} + \frac{1}{\rho}\frac{\partial \widehat{\tau}_{ij}}{\partial x_j} - \frac{\partial}{\partial x_j}(\widehat{u_i u_j} - \widehat{u}_i\widehat{u}_j) \tag{2.10}$$

where $\widehat{\tau}_{ij}$ is the stress tensor of the resolved field

$$\widehat{\tau}_{ij} = \nu \left(\frac{\partial \widehat{u}_i}{\partial x_j} + \frac{\partial \widehat{u}_j}{\partial x_i} \right) \tag{2.11}$$

and the last term of (2.10) is an unknown contribution related to the sub-grid scales. The LES approach requires a modeling of the term exact term τ_{ij}^S with the modelled one τ_{ij}^m as

$$\tau_{ij}^S = \widehat{u_i u_j} - \widehat{u}_i\widehat{u}_j \approx \tau_{ij}^m \tag{2.12}$$

which arises from the velocity scales that are smaller than the mesh size $\Delta^{''}$. From a physical point of view, τ_{ij}^m represents the dissipation of the turbulent kinetic energy associated with the sub-grid scales. Since the velocity field is no longer exactly evaluated but partly modeled the NSE in the framework of the LES approach read

$$\frac{\partial u_i^*}{\partial x_i} = 0 \tag{2.13}$$

$$\frac{\partial u_i^*}{\partial t} + \frac{\partial (u_i^*\, u_j^*)}{\partial x_j} = -\frac{1}{\rho}\frac{\partial p^*}{\partial x_i} + \frac{\partial \tau_{ij}^*}{\partial x_j} - \frac{\partial \tau_{ij}^m}{\partial x_j} \tag{2.14}$$

with $u_i^* \approx \hat{u}_i$. Several approaches have been proposed to model the contribution of the sub-grid scales depending on the investigated problem. Among the most common approaches, the one proposed by Smagorinsky (1964) who first derived the fundamental equations, and the dynamic model proposed by Germano et al. (1991) may be mentioned. For a detailed analysis of the LES approach and several applications, the reader should refer to Fröhlich (2006) or Sagaut (2006).

Reynolds-Averaged Navier-Stokes Equations

A different approach for the simulation of turbulent flow are the so-called Reynolds averaged Navier-Stokes Equations (RANSE), which allow evaluating only averaged flow features, not resolving the fluctuating behavior of the flow. This approach is based on the Reynolds average concept where the instantaneous fluid quantities are decomposed into a mean value and a fluctuating one. For the fluid velocity the Reynolds decomposition reads:

$$u_i = \langle u_i \rangle + u_i' \; . \tag{2.15}$$

The average process can be either in time, in one or more directions or an ensemble average, depending on the problem considered. Substituting (2.15) into (2.1) and (2.2) and an additional averaging of the resulting equations yields the RANSE

$$\frac{\partial \langle u_i \rangle}{\partial x_i} = 0 \tag{2.16}$$

$$\frac{\partial \langle u_i \rangle}{\partial t} + \frac{\partial (\langle u_i \rangle \langle u_j \rangle)}{\partial x_j} = -\frac{1}{\rho} \frac{\partial \langle p \rangle}{\partial x_i} + \frac{1}{\rho} \frac{\partial \langle \tau_{ij} \rangle}{\partial x_j} - \frac{\partial \langle u_i' u_j' \rangle}{\partial x_j} \; . \tag{2.17}$$

The last term does is not a function of the averaged quantities and needs, hence, to be modeled. It it commonly referred to as the Reynolds stress tensor, here represented by the symbol τ_{ij}^R and formally defined as

$$\tau_{ij}^R = \langle u_i' u_j' \rangle \tag{2.18}$$

althogh different from a stress by a factor $1/\rho$. The modeling of such term allows the solution of (2.17) and resolves the so-called "closure problem" of turbulence. Several approached have been proposed so far but here only a few are described, since a detailed description of the different methodologies goes far beyond the purposes of the present work.

Many modeling techniques involve a cardinal quantity in turbulence, the turbulent kinetic energy (TKE), represented by the symbol K. The TKE is defined by means of the fluctuating velocity as

$$K = \frac{1}{2} \langle u_i' u_i' \rangle \; . \tag{2.19}$$

Most of the modeling approaches base on the Boussinesq approximation, where τ_{ij}^R is set proportional to the trace-less mean strain rate tensor

$$\tau_{ij}^R \approx \nu_t \, \langle \tau_{ij} \rangle - \frac{2}{3} \, K \, \delta_{ij} \; , \tag{2.20}$$

where ν_t is the turbulence viscosity and δ_{ij} the Kroneker delta. After this assumption is made, a second procedure can be employed to evaluate ν_t, classified after the number of transport equations used: Zero-, one- and two equation models. In the zero-equation model the value of ν_t is determined without any transport equation i.e. algebrically, usually as the product of a length scale times a velocity scale. One example is the mixing-length-model, where

$$\nu_t = l_m \sqrt{\langle S_{ij} \rangle \langle S_{ij} \rangle} \tag{2.21}$$

where l_m is the so-called mixing length where

$$S_{ij} = \frac{1}{2} \left(\frac{\partial u_i}{\partial x_j} + \frac{\partial u_j}{\partial x_i} \right) \tag{2.22}$$

is the deformation tensor and its norm $\sqrt{\langle S_{ij} \rangle \langle S_{ij} \rangle}$ is used as velocity scale. The zero-equation model is not widely used and are employed only in very simple and didactic configurations. In one-equation-models, a transport equation is solved to account for the time and space variation of the turbulence. The transported quantity can be the turbulent viscosity ν_t, as proposed by Spalart and Allmaras (1994), or the TKE, which can be used to evaluate the time- and space-depending velocity scale, as proposed by Prandtl and Wieghardt (1947). Both approaches introduce some constants to fit the equation to the investigated flow and this restricts the use of such model for a limited number of cases. A similar approach is also employed in the framework of two-equation-models, where two transport equations are solved and used to evaluated the turbulent viscosity. The following models can be mentioned that belong to the most employed one:

- The $K - \epsilon$ -model, where an equation for K and an equation for the turbulent dissipation ϵ are solved and related to the turbulent viscosity as $\nu_t = c_\mu K^2/\epsilon$, with c_μ an empirical constant.

- The $K - \omega$ -model, where an equation for K and an equation for the turbulent frequency ω are solved and related to the turbulent viscosity as $\nu_t = K^2/\omega$.

- The shear stress transport (SST)-model, proposed by Menter (1994), which combines the two aforementioned models.

As mentioned, such models are based on the Boussinesq approximation and this implies the use of a unique value of ν_t in all flow direction, although variable in time and in space. This corresponds to assuming an isotropic turbulence behavior, where all flow directions are equivalent. As a consequence, the models listed above fail when the flow has a strong anisotropy, i.e. when there is at least one preferred direction. The Reynolds Stress Models (RSM) try to overcome this problem by solving six different transport equations, one for each component of τ_{ij}^R (which is a symmetric tensor) plus one additional equation, e.g. for ϵ. In the transport equations of the components of the Reynolds stress tensor terms arise which are related to the triple correlation of the fluctuating velocity field and these terms need also to be modeled, yielding an infinite number of equations for the exact evaluation of the field quantities. The closure problem of turbulence can thus be considered as the choice of the number of equations solved to determine the flow field, consisting in a compromise between accuracy and computational cost.

2.2   Numerical methods for two-phase flow simulation

As described in the previous section, several approaches can be employed for the simulation of single-phase flows. The possibilities rapidly multiply when a second phase is considered and this section briefly reviews the most common approaches for the simulation of bubbly flows. Here, the fluid is often referred to as the carrier phase and the bubbles as the disperse phase, since mainly disperse bubbly flows are considered. A sharp classification and distinction of the simulation techniques is an almost hopeless task, due to the strong interconnection of such methods and the constantly increasing number of algorithms. In this work, the description of the methods follows the level of accuracy that such methods employ: This arbitrary choice may not be the only one but it provides a clear path through the several approaches.

The first techniques considered are the ones that fully resolve the two-phase flow, i.e. that accurately describe the phase boundary and may be classified as "phase resolving" (PR) methods. Usually, the NSE are numerically solved for a single fluid on a fixed Cartesian grid and the several approaches differ from each other in the way the disperse phase is introduced in the governing equations.

Volume-Of-Fluid (VOF) This method consists in the advection of the interphase and in the evaluation of the mass fraction of fluid in each cell, associating a value of a scalar variable C to each cell: Usually $C = 1$ for cells completely in the fluid, $C = 0$ for cells completely in a bubble and $0 < C < 1$ for cells cut by the interphase. Proposed by Hirt and Nichols (1981) and by Youngs (1982), the main drawback is the difficult reconstruction of the interphase and much effort has been given to improve it: One of the most employed techniques is the Piecewise Linear Interphase Recalculation (PLIC), which bases on the assumption that the portion of interphase in each cell is defined as a plane with a arbitrary orientation. To account for the two phases, the NSE are solved with fluid properties, like viscosity and density, locally related to the value of C by means of weighted or harmonic averaging.

Front-tracking Method (FT) As the name suggests, this method consists in following or tracking the interphase in an explicit manner and was first proposed by Unverdi and Tryggvason (1992). This approach bases on the definition of the interphase as a number of connected markers which are advected by the fluid velocity. The interphase then needs to be reconstructed a posteriori after the new positions of the markers has been evaluated. This approach allows a precise and accurate evaluation of interphase-related quantities, such as the surface tension which is used to compute the additional force that is introduce in the NSE to account for the disperse phase. Several improvements have been developed so far, e.g. by Sint Annaland et al. (2006), Dijkhuizen et al. (2010) and Deen and Kuipers (2013) which upgraded the method to account also for heat transfer between the phases. The main disadvantage of this method is the computational cost, which is needed to constantly modify the mesh representing the bubble surface (the so called "re-meshing") to account for the bubble deformability.

Level-Set-Method (LS) Similar to the VOF, this technique defines a scalar variable, the level-set function φ, whose zero-level, i.e. $\varphi = 0$, defines the interphase between two fluids. Usually, $\varphi < 0$ is inside the bubble and $\varphi > 0$ inside the fluid. It was first introduced by Osher and Sethian (1988) but only later applied to two-phase simulations by Sussman et al. (1994). The similarity between LS and VOF consists in the direct advection of the interphase while the difference is that, for the VOF, the transition between the two fluids takes place over one cell, while for the LS the phase boundary is smoothed over a larger number of cells. Nevertheless, the strategy to solve the NSE is identical as in the VOF-method.

Improvements have been proposed by Sussman and Puckett (2000) for a coupled VOF-LS algorithm, by Hieber and Koumoutsakos (2005) for a Lagrangian LS method and by Enright et al. (2002) for a hybrid-particle-LS method.

Phase-field method (PF) For this method the phase boundary is smoothed over a given number of cells, leading to a phase boundary of finite width. The methods originates in the the phase-field theory proposed by Hohenberg and Halperin (1977) for phase transition and has been later developed as a numerical tool for the simulation of two-phase flows. It can be seen as LS method where the LS function is strictly related to the physics of the interface, usually to the mixing energy, as in (Yue et al., 2004). Its advantages lie in the physical definition of the PF function, which allows a more rigorous formulation from a physical point of view, regarding for example energy conservation and physical dynamics of the interface. Nevertheless, in the limit of vanishing boundary width, the PF method reduces to a sharp LS method. Eventually, this method requires adaptive quantities and is still very costly for the bubble simulations.

Euler-Lagrange-method A somewhat different approach, although similarities with the aforementioned methods may be found, is the Euler-Lagrange-method (ELM), where the disperse phase is explicitly tracked in a Lagrangian manner. Several degrees of accuracy and different types of coupling between the objects and the fluid can be employed, but some common features can be found. In this framework, each disperse object is tracked separately and its trajectory is evaluated by integrating the equation of motion, usually in the form:

$$m_p \frac{\partial \mathbf{u}_p}{\partial t} = \mathbf{F}_p \tag{2.23}$$

where m_p is the mass of the particle, $\mathbf{u}_p$ the particle velocity to be computed and $\mathbf{F}_p$ the sum of all forces acting on the particle[1]. The drag force, usually is the most important one and thus considered in all approaches. Which forces are additionally considered defines the degree of accuracy of the method. Among these range the lift force, the pressure force, the virtual mass force, the history force, etc (Crowe, 2005). Usually, each force is evaluated by using specific models which rely on empiric relations and/or coefficients. Among the ELM, the following classification can be established, according to an increase in the total void fraction considered.

- One-way coupling: The fluid influences the trajectory of the particles but the particle does not induce any modification of the fluid field.

- Two-way coupling: The fluid influences the particle *and* the particle influences locally the fluid. This is usually accomplished by introducing in the NSE an additional force, namely $\mathbf{F}_f = -\mathbf{F}_p$.

- Three-way coupling: The influence of each particle is felt not only by the fluid itself, but also the by other particles in an indirect manner, e.g. through wake effects.

- Four-way coupling: Particles are influenced by other particles also in a direct manner, i.e. considering particle-particle-collisions.

---

[1]In some methods, also a simular equation is solved for the particle rotation, but rarely.

Such methods are mainly employed in two-phase flows where the particles are very small, i.e. comparable with the smallest spatial scales in the flow. In this context, the point-particle approximation, i.e. considering the particles as mass points, has proven to provide valuable results. Nevertheless, the accuracy is highly reduced with respect to the aforementioned PR methods, since the particle equation of motions employes empirical correlation which may not be able to capture all flow phenomena.

The Immersed Boundary Method (IBM) can be considered at the border between the PR and the ELM approaches since it presents the features of both approaches. As a PR method, it fully describes the geometry of the immersed objects and, as the ELM, it tracks the motion of each object in a Lagrangian manner. The advantages with respect to an ELM is that the forces acting on the objects are directly evaluated by the surrounding fluid flows and no empirical correlation is needed for the particle equation of motion. The IBM was first proposed by Peskin (1977) and subsequent improvements have been reviewed by Mittal and Iaccarino (2005). An efficient version of the IBM, inspired by Uhlmann (2005), was developed by Kempe and Fröhlich (2012a) and is employed here to perform simulations of bubbly flows. A description of the method proposed in (Kempe and Fröhlich, 2012a) and successive improvements to account for the motion of bubbles are reported in the following section.

2.3   Immersed Boundary Method

In the framework of the IBM employed in the present work the bubbles are introduced in the NSE by means of additional forces, hence (2.2) is modified to

$$\frac{\partial u_i}{\partial t} + \frac{\partial \left(u_i\, u_j\right)}{\partial x_j} = -\frac{1}{\rho}\frac{\partial p}{\partial x_i} + \frac{1}{\rho}\frac{\partial \tau_{ij}}{\partial x_j} + f_i + f_{I,i}\,, \tag{2.24}$$

where f_i collects the volume forces and $f_{I,i}$ represents the IBM forces in the i direction. The spatial discretization, employing a Finite Volume Method, is based on an equidistant Cartesian grid with a staggered arrangement, with central second-order schemes for convection and diffusion. The time advancement of second order is performed by a predictor-corrector approach consisting of an explicit three-step Runge-Kutta scheme for the convective terms and a semi-implicit Crank-Nicholson scheme for the diffusive terms, completed by a pressure correction equation. The coupling between the disperse phase and the fluid motion is performed by means of a direct forcing approach as proposed by Mohd-Yusof (1997). For this purpose, each immersed object is described by a given number of forcing points on the phase boundary, at which the coupling between fluid and disperse phase is realized. The interpolation of Eulerian velocities and the successive spreading of Lagrangian forces is accomplished via a weighted sum of discrete three-point Dirac delta functions as proposed by Roma et al. (1999).

The dynamics of each element of the disperse phase is described by the position of the center point $x_{p,i} = \mathbf{x}_p$ and its derivative in time, the translational velocity $\mathbf{u}_p$, as well as the angular position together with the rotation rate $\boldsymbol{\omega}_p$. For clarity, bold symbols are here employed to represent vectors collecting the components in the three Cartesian direction. In the present study, only spherical bubbles are considered and the particle equations of motion for the translational and the angular velocities read

$$m_p \frac{d\,\mathbf{u}_p}{d\,t} = \rho \int_\Omega \boldsymbol{\tau} \cdot \mathbf{n}\,dS \;+\; (\rho_p - \rho)\,V_p\,\boldsymbol{g} + \mathbf{F}_s \tag{2.25}$$

$$I_p \frac{d\,\boldsymbol{\omega}_p}{d\,t} = \rho \int_\Omega \mathbf{r} \times (\boldsymbol{\tau} \cdot \mathbf{n})\,dS \;+\; \mathbf{M}_s\;, \tag{2.26}$$

where Ω is the particle surface, $I_p = 2/5\; m_p\; r_p^2$ the moment of inertia of the spherical particle, r_p the particle radius and $\mathbf{r}$ the vector from the particle center to a point on the particle surface. Additional source terms are denoted by $\mathbf{F}_s$ and $\mathbf{M}_s$, respectively. In this work, these sources arise only from bubble-bubble and bubble-wall collision events, as explained below. The temporal discretization of (2.25) and (2.26) for very light objects, such as bubbles, faces the problem that m_p and I_p become very small. A dedicated time scheme was hence developed in (Schwarz et al., 2015) to cope with this situation. It is based on a particular definition of a numerical virtual mass which yields an efficient, stable and versatile algorithm.

Bubble-bubble and bubble-wall collisions are accounted for by a repulsive potential force depending on the distance vector between the respective surfaces. The force induced on bubble i by bubble j reads

$$\mathbf{F}_{i,j} = \begin{cases} K\,(\varphi_n - d_s)\;\mathbf{d}_{i,j} \,/\, ||\mathbf{d}_{i,j}||, & \text{if } \varphi_n \le d_s, \\ 0, & \text{else}\;, \end{cases} \tag{2.27}$$

where $\mathbf{d}_{i,j} = \mathbf{x}_{p,j} - \mathbf{x}_{p,i}$ is the distance vector between the bubble centers. The quantity $\varphi_n = ||\mathbf{d}_{i,j}|| - r_{p,i} - r_{p,j}$ represents the distance between the two bubble surfaces, and $r_{p,i}$ and $r_{p,j}$ the radii of the respective bubbles. The quantity d_s in (2.27) is a safety clearance to prevent bubbles from overlapping and is chosen equal to the mesh width Δ here. The model parameter K in (2.27) is determined according to the surface tension of the colliding objects, setting $K = 2\pi\sigma$, with σ the surface tension. In the present study, the Eötvos number $Eo = |\mathbf{g}|\; d_p^2\; (\rho - \rho_p)/\sigma$ was set to 0.3, which is a realistic value for bubbles of diameter of $1\; mm$ in contaminated water. Selecting the bubble diameter, this yields the surface tension σ which is then used for the collision model. The resulting value of about $30mN/m$ is somewhat smaller than measured, but as the shape of the bubbles is fixed the value of σ only enters in the collision model. Collisions, on the other hand, play a negligible role in the configuration investigated as demonstrated in Sec. 3.3.4 below, so that the actual value of σ is practically irrelevant. Nevertheless, with this value, bubble-wall and bubble-bubble collisions take place during about 10 time steps, for physical reasons, so that they are appropriately resolved whit the time step dictated by the CFL criterion[1].

The in-house code PRIME, into which the above method was implemented, was extensively validated for different configurations such as a fixed sphere in cross flow (Kempe and Fröhlich, 2012a) and settling and rising spheres at finite Reynolds number (Schwarz and Fröhlich, 2014). Additionally, the code was employed and validated for the simulation of sediment transport (Kempe et al., 2014), close-packed structures of spheres (Heitkam et al., 2012), and the motion of bubbles in a magnetic field (Schwarz and Fröhlich, 2014).

---

[1]See (Kempe and Fröhlich, 2012b) for a different collision model designed for particles where the physical conditions are much stiffer.

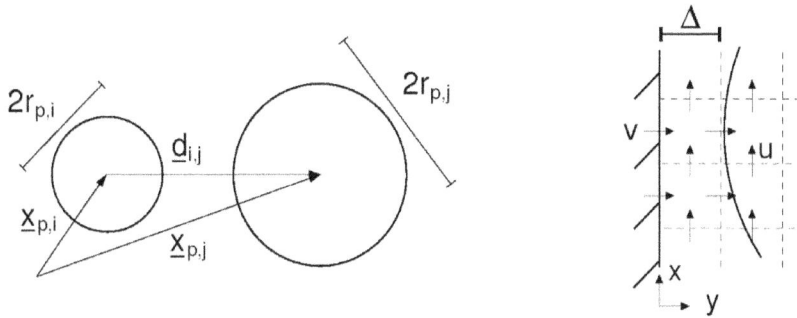

Figure 2.1: Illustration related to the collision modeling employed. Left: Geometrical quantities entering the collision model. Right: Situation where during a bubble-wall collision the bubble is closest to the wall. The figure highlights that there is always at least one Eulerian point available to compute the wall-shear stress.

3 Configuration and reference case

The present chapter consist of three parts. In the first, Sec. 3.1, the configuration investigated is introduced and the complete set of simulation performed is presented. Afterward, the numerical approach introduced in Sec. 2.3 is validated for the chosen parameter range for small bubbles by means of two reduced problem. Finally, Sec. 3.3 is devoted to the analysis of the the reference case *SmMany*, where a large number of bubbles is considered. Some of the results reported in this chapter have been presented in Santarelli and Fröhlich (2014).

3.1 Numerical setup

The simulations reported here were performed in for channel flow configuration representing the flow between two parallel vertical walls at distance H. On both walls a no-slip condition for the fluid velocity was applied, with y being the wall-normal coordinate. The flow is periodic in the streamwise (x) and spanwise (z) direction. The gravity force acts in the negative x-direction and the mass flow rate was kept constant by means of an instantaneously adjusted volume force in the x-direction, f_x. Hence, the bulk velocity U_b, averaged over liquid and gas was constant in time and set equal for all simulations. This yields the same bulk Reynolds number $Re_b = U_b H/\nu$ for all simulations. The value of Re_b was set to 5263, so that turbulence, even if low, is sustained also in the unladen case. Figure 3.1 portrays an instantaneous snapshot of the two-phase mixture in the *SmMany* case and elucidates the configuration investigated.

The choice of the channel extension in the streamwise direction is crucial since large-scale flow features may have a strong influence on the small-flow features, as reported in Sec. 1.1. It has been observed in several studies of the literature (Lance and Bataille, 1991; Panidis and Papailiou, 2000; Hosokawa and Tomiyama, 2004) as well as in own preliminary tests that bubbles in the present regime enhance turbulence and reduce correlation lengths. For efficiency, the size in the streamwise direction, L_x, was hence selected somewhat smaller than the canonical size of a single-phase turbulent channel flow, e.g. in (Kim et al., 1987). As a consequence, turbulent quantities in the unladen case such as correlation functions, Reynolds stresses, etc., are not expected to match exactly when simulating unladen flow in the present geometry. This was accepted since not the single-phase flow but the bubble-laden configuration is the target here. Nevertheless, good agreement was found for first and second order statistics when compared with (Kim et al., 1987).

The disperse phase increases the level of turbulence substantially and yields a very different situation. The present bubble-laden channel is comparable to the one investigated by Uhlmann (2008), laden with heavy particles. In a later study, the same configuration was investigated in a channel of twice the extent in the streamwise direction (Garcia-Villalba et al.,

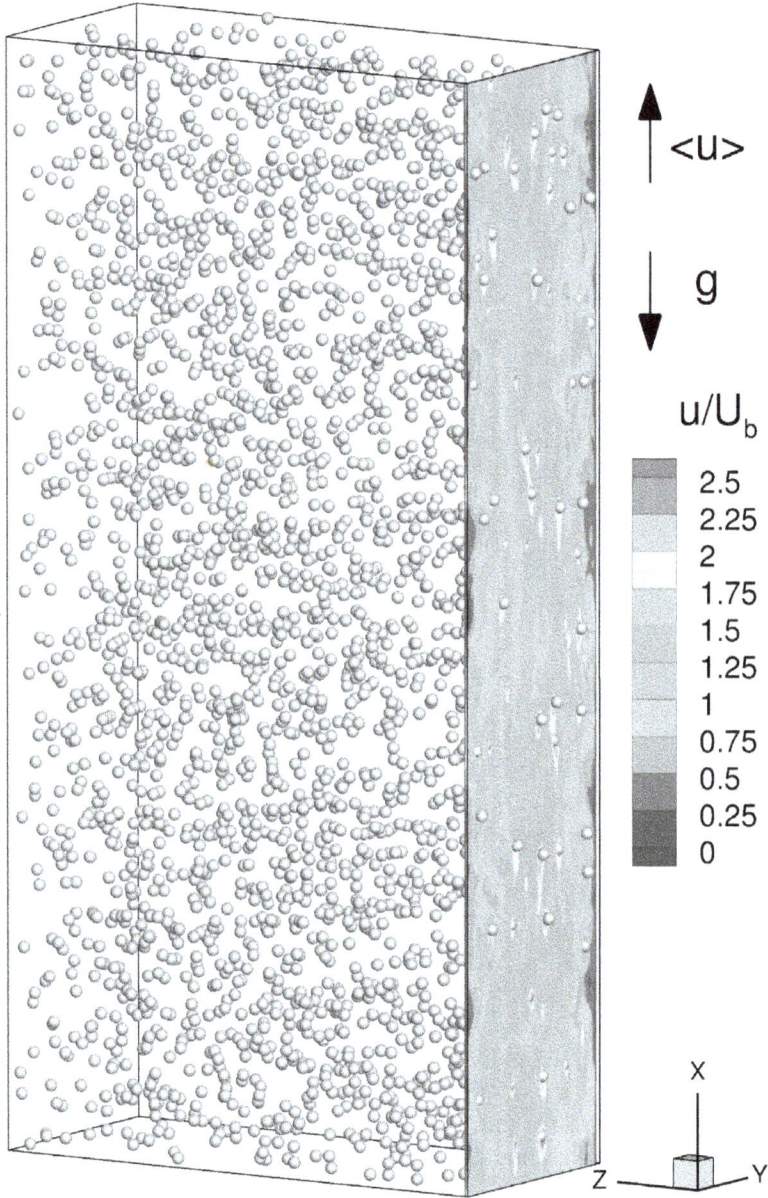

Figure 3.1: Computational domain for the flow investigated. The channel walls are positioned at $y = 0$ and $y = H$, while the flow is periodic in the x and z direction. The bubble size is to scale and the vertical plane with the contour plot shows the streamwise fluid velocity component. The snapshot is taken for an arbitrary instant in time during the simulation *SmMany*.

2012). These authors did not find any significant difference regarding Eulerian statistics and particle statistics, except for the correlation function in streamwise direction, as expected.

Reference	Physics	L_x/H	L_z/H	Re_b	ϕ_{tot}
Kim et al. (1987)	unladen flow	6.28	2.09	6600	[-]
Uhlmann (2008)	particulate flow	4.00	2.00	5400	0.42%
Garcia-Villalba et al. (2012)	particulate flow	8.00	2.00	5400	0.42%
Lu and Tryggvason (2013)	bubbly flow	1.57	0.78	$Re_\tau = 250$	3.00%
present	bubbly flow	4.41	2.21	5263	see Table 3.2

Table 3.1: Selected quantities of the present studies compared to the literature. The shear Reynolds number Re_τ is reported in Tab. 3.3 below for comparison. In (Lu and Tryggvason, 2013) the bulk Reynolds number for the two-phase flow is not given and, from Fig. 3(b) therein, $Re_b \approx 4000$ can be estimated.

Lu and Tryggvason (2013) performed DNS of bubble-laden channel flow whose dimensions are much smaller than the usual ones for channel flow. To the best of the author's knowledge, so far no DNS of phase-resolved bubbles has been performed in a periodic channel equal or larger in size than the one investigated in the present work. Table 3.1 summarizes the channel dimensions for selected simulations from the literature to which the present results will be compared. The extension of the channel investigated in the present work are $4.413H \times H \times 2.207H$ in streamwise, wall-normal and spanwise direction, respectively. The Cartesian grid employed has the same step size $\Delta = H/232$ in all Cartesian directions resulting in $1024 \times 232 \times 512$ grid points in the x-, y- and z-direction, respectively. This leads to around 120 Million grid points in total.

In the present work, bubbles are simulated as spherical objects of fixed shape which is justified by the moderate value of the bubble Reynolds number Re_p and by the the small value of the Eötvös number considered here (Clift et al., 2005). The density of the bubbles is much lower than the density of the fluid and was set to $\rho_p/\rho = 0.001$. A no-slip condition between fluid and bubble velocity was applied at the phase boundary, representing the case of air bubble rising in contaminated water. This is backed by the following numerical simulations and experimental investigations. Tasoglu et al. (2008) performed numerical simulations of a rising bubble in quiescent fluid with surfactant and found that the surface velocity of the contaminated bubble nearly vanishes and that the bubble behaves similarly to a solid sphere when it reaches a steady motion. In the same work a rigidifying effect caused by surfactants was also observed, which further justifies to model bubbles as rigid spheres. Tagawa et al. (2010) experimentally investigated rising bubbles in water contaminated with different surfactants and found that contaminated bubbles behave like rigid particles in water, exhibiting a drag coefficient which is almost the same as that of rigid particles. Numerically, Aland et al. (2013) discussed the correctness of a no-slip condition for bubbles rising in liquid metal. They compared two different numerical approaches, the IBM employed here and a Navier-Stokes-Cahn-Hilliard model. The simulations reported therein show that, when the surface tension is very large, the bubble shape remains constant and the contaminants at the surface result in an effective no-slip condition.

Five simulations of bubble-laden flows were performed in the present study:

- Simulation *SmMany* addresses a dense swarm of 2880 small bubbles with a total void fraction of 2.14%.

- Simulation *SmFew* addresses a dilute swarm of 384 small bubbles of the same size as in *SmMany* and allows investigating the influence of the total void fraction on the flow.

- Simulation *LaMany* addresses a dense swarm of 913 large bubbles with the same void fraction as the *SmMany* case. It allows investigating the influence of the bubble size.

- Simulation *BiDisp* addresses a dense bidisperse swarm of 1896 bubbles, where half of the void fraction is made of small bubbles and the other half of large bubbles, with the same void fraction as the *SmMany* and *LaMany* cases. This case allows investigating the influence of the bidispersity of the the swarm.

- Simulation *SmManyDo* is performed under the same conditions as in the *SmMany* case in a downward flow configuration, where the flow direction and the gravity vector act in the same direction and, with this case, the influence of the flow direction is investigated.

The parameters of all simulations are collected in Tab. 3.2; All other physical and numerical parameters were identical in all simulations. As a reference, an additional simulation was conducted under exactly the same conditions without any bubbles, labeled *Unladen* in the following.

Simulation	*SmMany*	*SmFew*	*LaMany*	*BiDisp*			*SmManyDo*
					SM	LA	
d_p/H	0.052	0.052	0.075		0.052	0.075	0.052
N_p	2880	384	913	1896	1440	456	2880
d_p/Δ	12.1	12.1	17.4		12.1	17.4	12.1
ϕ_{tot}	2.14%	0.29%	2.14%	2.14%	1.07%	1.07%	2.14%
$\mathbf{g} \cdot \langle \mathbf{u} \rangle$	<0	<0	<0	<0			>0

Table 3.2: Parameters for the channel flow simulations.

A mixture can be considered dilute if the void fraction ϕ_{tot} is below 2%, which is about the void fraction in the denser swarms considered here. Even for a total volume fraction up to 6.5%, i.e. for intermediate mixtures, collisions are not significant and are not supposed to play a significant role (Michaelides, 2006).

The following procedure to determine statistical quantities was employed for the bubble-laden simulations. Bubbles were introduced in the fully turbulent unladen channel flow at $t = t_{p,0}$. The initial position of the bubbles was selected at random and the initial velocity set equal to the velocity of the Eulerian grid point closest to the bubble center. Twenty bulk time units $20T_b$, with $T_b = H/U_b$, were regarded as the initial transient time, during which target quantities, such as the automatically adjusted volume force f_x and the wall-shear stress τ_w, indeed, reached a stationary state. Afterward statistics for fluid and disperse phase were collected over $400\ T_b$. This value is higher by a factor of 7 compared to (Uhlmann, 2008) and by a factor of around 9 compared to (Garcia-Villalba et al., 2012). For the simulation *SmFew* this was increased to $500\ T_b$ for better convergence. Whenever possible, additional averaging in x- and z-direction was employed. The time history of the shear Reynolds number for the *Unladen*, the *SmMany* and the *SmFew* cases is depicted in Fig. 3.16 below for illustration.

3.2 Simulation of single bubble cases

Before investigating the dynamics of bubble swarms, two reduced problems will be addressed
in this section to validate the numerical approach and to provide reference data for the subsequent physical analysis of the swarm simulations. Section 3.2.1 deals with the simulation of a
single bubble rising in quiescent fluid with a bubble Reynolds number close to the one of the
small bubbles in the channel flow. This simulation was performed to validate the numerical
method in the considered parameter regime and to obtain reference data with exactly the
same method for comparison with turbulent flows. In Sec. 3.2.2 the lift force acting upon a
fixed bubble in cross-flow with constant shear rate is investigated, since the shear-induced lift
force is supposed to play a significant role in the dynamics of bubbles in turbulent channel
flow. As in the case of the bubble rising in quiescent fluid, the bubble Reynolds number was
chosen close to the one of small bubbles in the channel, to perform an additional validation of
the numerical scheme for the investigated parameter range. Analogous problems addressing
large bubbles will be provided in Sec. 5.1 below.

3.2.1 Single bubble in quiescent fluid

The computational domain is a rectangular vertical column with extension of $175.6\,d_p \times$
$16.6\,d_p \times 16.6\,d_p$ in the upward and in the two horizontal directions, respectively. To approach
the flow of a bubble in an unbounded flow, the domain must be large enough in order to
have no influence on the bubble dynamics. In (Schwarz and Fröhlich, 2014) a domain of
extensions $30d_p \times 6d_p \times 6d_p$ with periodic conditions was used for a similar case, and a
detailed justification was provided that this size is sufficient. In the present simulation a
free-slip condition is used at the lateral boundaries, and to be on the safe side an even larger
domain was chosen so that the configuration models an unbounded domain without any
doubt.
Reference quantities t_{ref} and u_{ref} are defined as follows:

$$t_{ref} = \sqrt{\frac{d_p}{g\,|\pi_\rho - 1|}} \quad , \quad u_{ref} = \sqrt{d_p\,g|\pi_\rho - 1|} \quad , \tag{3.1}$$

and the density ratio $\pi_\rho = \rho_p/\rho$ is equal to 0.001. This case is labeled *SingleSm* in the
following.
Based on the terminal rise velocity $u_{p,\infty}$ obtained from averaging over a duration of $100t_{ref}$,
the bubble Reynolds number was determined as

$$Re_{p,\infty} = \frac{u_{p,\infty}\,d_p}{\nu} \quad . \tag{3.2}$$

For the present simulation its value is around 265.7, which is comparable to the ones of
bubbles rising in the channel flow addressed later on.
Several investigations have been performed in the literature to elucidate the different regimes
of a rising sphere in quiescent fluid. As extensively reviewed by Ern et al. (2012), different
types of motion are possible, strongly dependent on the Reynolds number and the density
ratio: a straight path, periodic oscillations, and chaotic motion, for example. Here, the own
results are compared with the work of Horowitz and Williamson (2010), who experimentally
investigated the rise and fall of a sphere, varying $Re_{p,\infty}$ and π_ρ. As mentioned in Sec.

3.1 above, bubbles are simulated as spherical objects and a no-slip condition is applied at the phase boundary between bubble and fluid so that a comparison with a rigid sphere is appropriate. According to (Horowitz and Williamson, 2010), the bubble path is expected to present a zigzag shape with the present set of parameters. For the smallest density ratio investigated in that reference, $\pi_\rho = 0.07$, the threshold between straight oblique path and zigzag path is $Re_{p,\infty} = 260$, which is very close to the one of the present simulation. As illustrated in Fig. 3.2 the bubble path is almost straight, since the final horizontal displacement from the starting position is less than $2d_p$, while the bubble traveled over a vertical distance of 170 d_p. Nevertheless, a very slight periodicity can be noticed in the horizontal bubble velocity components v_p and w_p shown in Fig. 3.3, left. This was confirmed by various flow visualizations, and an example is provided in Fig. 3.3, right, showing the instantaneous vortex structures in the wake of the bubble by means of a λ_2-isosurface. It shows the presence of both, counter-rotating vortices far from the bubble related to a straight path, as well as the shedding of bent, spiralling vortices in the middle of the wake related to the zigzag path. For the considered Reynolds number the bubble wake hence exhibits both features of a straight and of a zigzag path and the trajectory demonstrates that this case is located at the border of the two types of motion.

After the initial transient time of around $15t_{ref}$ which correspond to $L_p/d_p \approx 20$, the bubble

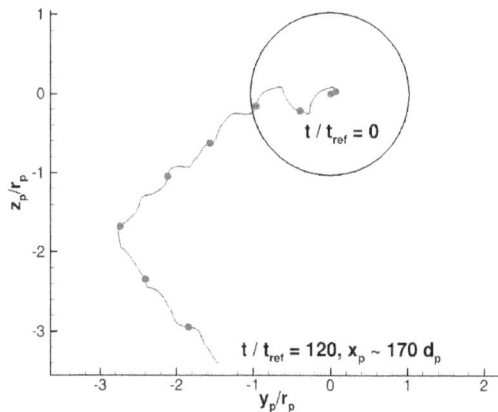

Figure 3.2: Top-view (zy) of bubble path for case *SingleSm* indicated by a continuous line. The size of the bubble is marked by a circle around the start of the trajectory at $t/t_{ref} = 0$. The end of the recorded trajectory is reached at a vertical position of about $170d_p$ after a time of $120t_{ref}$.

has reached a steady state and rises at nearly constant velocity, as visible in Fig. 3.3, left. In their experimental investigation of light and heavy spheres in quiescent fluid, Horowitz and Williamson (2010) state that "The terminal velocity U was typically reached by a distance $L_p/d_p \approx 8$ after being launched, except for spheres exhibiting an initial transient motion", which reached a steady state after $L_p/d_p \approx 20$. In a different study, Jenny et al. (2004) performed numerical simulations to investigate, among others, the rise of light spheres in quiescent fluid. For a density ratio $\rho_p/\rho = 0.5$ and comparable $Re_{p,\infty}$, the transition from the vertical to the steady, oblique path occurs at $t/t_{ref} \approx 20$. This value is comparable with the one where the bubble reaches its steady behavior in the present simulation. Hence, the present simulation is in good agreement with the observations reported in the literature and, according to the references, no additional transition is to be expected at later times for this situation. Based on $u_{p,\infty}$, the drag coefficient

$$C_D = \frac{4}{3} \frac{d_p \left| \rho_p - \rho \right| g}{\rho \, u_{p,\infty}^2} \,, \tag{3.3}$$

was evaluated, yielding $C_D = 0.75$. This value is in very good agreement with (Horowitz and Williamson, 2010), as illustrated in Fig. 3.4. This Figure again highlights the fact that the present case is located at the border of the regimes of straight and oscillatory motion.

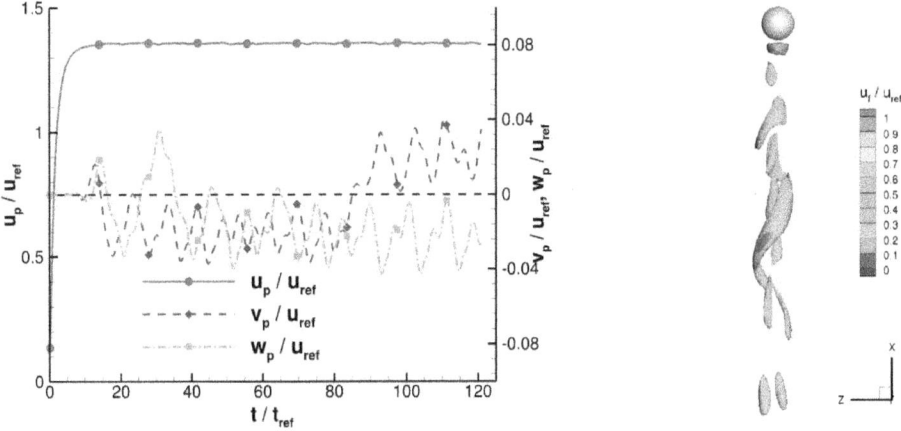

Figure 3.3: Left: Time history of bubble velocity components for the case *SingleSm*. The straight dashed line represents the zero value of the right ordinate. Right: Flow visualization for case *SingleSm*: Isosurface of λ_2, colored with the instantaneous fluid velocity in the x-direction. The λ_2 isosurface was chopped at the rear end of the bubble not to cover the bubble itself.

3.2.2   Fixed bubble in shear flow

When bubbles rise in a turbulent channel flow, two main aspects differ from the case of a rising bubble in quiescent flow previously investigated: the turbulence of the flow and the mean shear of the fluid velocity in the wall-normal direction. Ambient turbulence is expected to modify the bubble dynamics in a rather stochastic manner. The shear of the mean flow, instead, induces a mean lift force on the bubble and is supposed to play an important role in the dynamics of the bubbles in the channel flow. Therefore, the lift force induced is now analyzed in a slightly simpler situation, a fixed bubble in shear flow. This analysis provides qualitative information on the role of the lift force that will later on be referred to for the analysis of the bubble dynamics in the channel flow.

Many investigations have been performed to analyze the wake of a fixed sphere in uniform cross flow and in cross flow with a shear rate. As for an ascending sphere, different types of wakes are observed, depending on the Reynolds number: Steady axisymmetric wake, steady wake with two counter-rotating vortexes, periodic shedding of vortices and chaotic wake. Bouchet et al. (2006) investigated the different types of wakes behind a fixed sphere in uniform cross flow at moderate Reynolds number $Re_{p,c} = u_c \, d_p / \nu$, based on the freestream

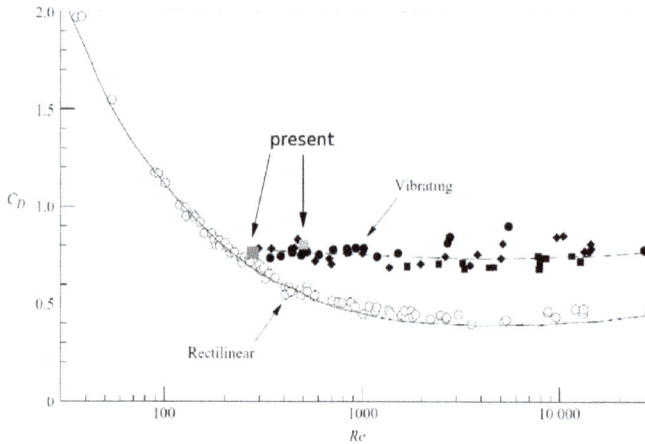

Figure 3.4: Comparison of drag coefficient for rising and falling spheres: Result of simulation *SingleSm* marked by a red square at $Re_p = 265$ in a picture taken from (Horowitz and Williamson 2010). Black symbols: vibrating sphere, with non-straight paths, obtained with various density ratios. Empty symbols: rectilinear trajectory. Solid line: drag for stationary sphere according to correlation of Wieselsberger (1921) and Liebster (1927). Dashed line: interpolating curve of experimental results. Results for *SingleLa* (green square at $Re_p \approx 500$) will be discussed in Sec. 5.1.1 below. Usage of Figure with permission from Cambridge University Press.

velocity u_c at the location of the particle center. For $Re_{p,c} = 260$ the wake was found to be stationary but non-axisymmetric, inducing a lift force on the particle, even in the absence of a velocity gradient.

The findings of the pioneering work of Kurose and Komori (1999), devoted to the lift force experienced by a fixed sphere in a constant shear flow, can be summarized as follows. For $Re_{p,c} < 60$, the shear-induced lift force is always directed toward the side of the sphere where the relative fluid velocity is higher. For $Re_{p,c} > 60$, instead, which is the range corresponding to the bubbles in the present channel flow, the sphere experiences a lift force toward the lower relative velocity, independent of the shear rate. The magnitude of the lift force increases, for $Re_{p,c} > 60$, both with increasing $Re_{p,c}$ (at fixed shear) and increasing shear (at fixed $Re_{p,c}$). An additional contribution regarding the flow around a sphere in shear flow was provided by Bagchi and Balachandar (2002), but their analysis was limited to $Re_{p,c} \leq 200$, while the Reynolds number of the bubbles in the channel investigated below is between 230 and 260. Nevertheless, for $Re_{p,c} = 200$, the negative lift coefficient found in (Kurose and Komori, 1999) was confirmed. Furthermore, these authors investigated the flow around spheres that were free to rotate and found practically no influence of the free rotation on the lift force for $Re_{p,c} = 200$. This result suggests that for the investigated regime the lift on a free rotating bubble does not differ significantly from the lift on a fixed one.

For validation purposes and to gain reference data, a fixed bubble in shear flow was simulated in the present work and the simulation is labeled *FixedSm*. The computational domain was defined as $[0; 25d_p] \times [0; 12.5d_p] \times [0; 12.5d_p]$ in the streamwise, normal and spanwise direction, i.e. x-, y-, and z-direction, respectively. At the inlet a linear velocity profile was imposed, with non-dimensional shear parameter $S = (\partial u / \partial y) r_p / u_c = 0.1$, where u_c is the freestream velocity at midspan. The sphere was positioned at $(x/d_p, y/d_p, z/d_p) = (6.25, 6.25, 6.25)$ and the Reynolds number $Re_{p,c}$ set to 260 for the reason mentioned above. The ratio between diameter and step size of the mesh was 12, as in the channel flow simulations below, yielding a total number of grid points of around 6.7 Million. As demonstrated by the results obtained, the resolution is adequate to capture the principal features of the flow, like wake structures and global forces on the sphere, as will be now discussed.

Figure 3.5 shows a lateral view and a side view of the sphere and of the wake structures represented by isosurfaces of the vorticity in the streamwise direction and colored by the z-component of the fluid velocity, after a steady stated was reached. The wake is symmetric with respect to the xy-plane cutting the bubble center but a slight asymmetry is observed with respect to the xz-plane due to the imposed velocity shear. A slight asymmetry of the wake was also found in (Bouchet et al., 2006) for the same $Re_{p,c}$ in uniform flow. Additionally, the history of the force coefficient in the y-direction was evaluated as

$$C_y = \frac{F_y}{1/2\, \rho\, \pi\, u_c^2\, (d_p/2)^2} \, . \tag{3.4}$$

As expected, the wake is stationary and C_y is constant in time and always negative, indicating that the lift force is directed toward the side of the sphere with the lower relative velocity, throughout, in accordance with the results of Kurose and Komori (1999), as portrayed in Fig. 3.6: The value of $C_y = -0.058$ is in good agreement with the one provided in (Kurose and Komori, 1999).

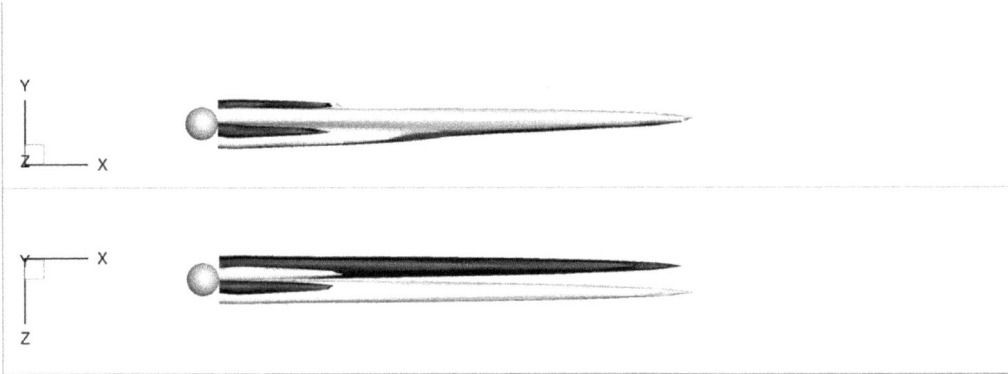

Figure 3.5: Wake structure behind a fixed sphere in shear flow for simulation *FixedSm*. The wake is represented by isosurfaces of the vorticity in the streamwise direction $\omega_x = \pm 2.5 U_c/d_p$, colored with the instantaneous fluid velocity in the z-direction. The velocity gradient is oriented in the y-direction. The vorticity isosurfaces are chopped at the rear end of the bubble not to cover the bubble itself. Top: Side view. Bottom: Top view.

In which direction is the lift force directed for a bubble rising in a channel flow? If a bubble rises in the channel, the rise velocity is higher than the one of the fluid, yielding a positive relative velocity. Due to the mean shear, the two sides of the bubble experience a different relative velocity: it is lower at the side of the bubble toward the channel center, where the fluid velocity is higher, and higher for at side of the bubble toward the wall, where the fluid velocity is lower. As discussed above, the lift force for the chose bubble

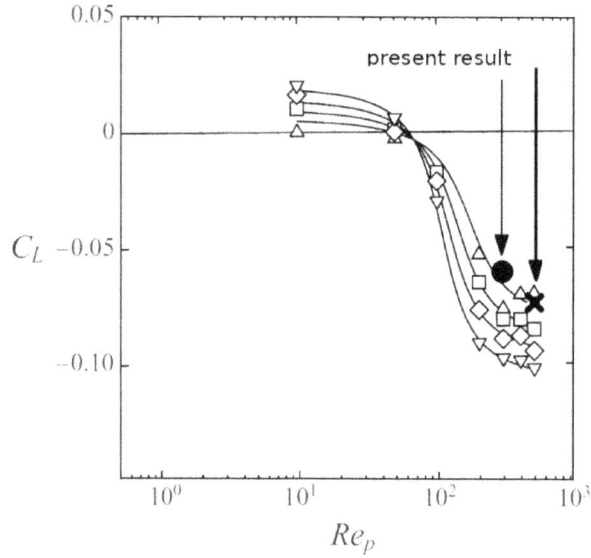

Figure 3.6: Comparison between numerical results of Kurose and Komori (1999) and the *FixedSm* simulation, black circle. Results for *FixedLa* (black cross at $Re_p \approx 500$) will be discussed in Sec. 5.1.1 below. Usage of Figure with permission from Cambridge University Press.

Reynolds number regime is directed toward the side of bubble that experiences the lower relative velocity: This yields a lift force directed toward the channel center moving the bubble in this direction. Additional mechanisms that influence the wall-normal distribution of bubbles, like turbophoresis and collisions, will be discussed with the swarm simulations presented below.

3.3   Results for the case *SmMany*

This section is devoted to the analysis of the denser swarm of small bubbles, case *SmMany*, used as reference case. The analysis starts with flow visualizations and a description of fluid structures, qualitatively and quantitatively. Subsequently fluid statistics and disperse-phase statistics are addressed.

3.3.1   Fluid phase: Flow structures and correlation functions

A first qualitative impression of the flow field is given by the representation of three-dimensional fluctuations of the streamwise velocity u'_{3d}

$$u'_{3d}(x, y, z, t) = u(x, y, z, t) - \langle u \rangle_{xz}(y, t) \tag{3.5}$$

where the symbol $\langle ... \rangle_{xz}$ represents averaging over dimensions indicated in subscript. Fluid quantities are averaged with the interior of the bubble being excluded.

In the case of the unladen channel flow, shown in Fig. 3.7, left, the typical elongated structures in streamwise direction can be found, mostly in the near wall regions. These structures consist of regions of positive and negative fluctuations, alternating in spanwise direction. When a sizable number of bubbles is introduced the picture is quite different, as depicted in Fig. 3.7, right. The instantaneous flow is dominated by long streamwise regions of negative fluctuations and these structures are rougher compared to the ones in the unladen channel flow, due to the local perturbations introduced by the bubbles. Furthermore, the magnitude of these perturbations is larger, as reflected by the different scale required to represent them with isosurface and contour plots. Regions of positive fluctuations (not shown here) are much shorter and of size $\mathcal{O}(d_p)$, as they are strongly related to individual bubbles and generated by bubble wakes. Visualization of velocity fields, stored at high frequency, allowed to investigate the temporal behavior of the longitudinal structures. The time interval of such visualization was limited to around $6T_b$ due to the size of the data stored amounting to around 4GB for each instantaneous flow field. The elongated structures persist over the whole visualization time and rise in streamwise direction with a velocity of around U_b.

A quantification of the streamwise turbulent length scale is possible by means of the two-point correlation function of the velocity fluctuations in the streamwise direction

$$R_{uu}(y, \Delta x) = \frac{\langle u'(x, y, z, t)\, u'(x + \Delta x, y, z, t) \rangle_{xzt}}{\langle u'^2(x, y, z, t) \rangle_{xzt}}, \tag{3.6}$$

where $u'(x, y, z, t) = u(x, y, z, t) - \langle u \rangle_{xzt}$. It is shown in Fig. 3.8 and 3.9 for the near-wall region and for the center region, respectively, and exhibits substantial differences between bubble-laden and unladen flow.

On small scales, bubbles generate turbulent fluctuations and the correlation is therefore reduced, yielding a reduction of turbulent length scales. This holds for both regions investigated, even if this feature is more pronounced near the wall. A reduction of the velocity correlation for small scales was also found by Lance and Bataille (1991) and by Panidis and Papailiou (2000), who performed experiments of bubbly flows in grid-generated turbulence. In these papers, the bubbles were found to mainly influence the small scales, and the reduction of the correlation was higher for denser swarms, i.e. for higher void fraction. The

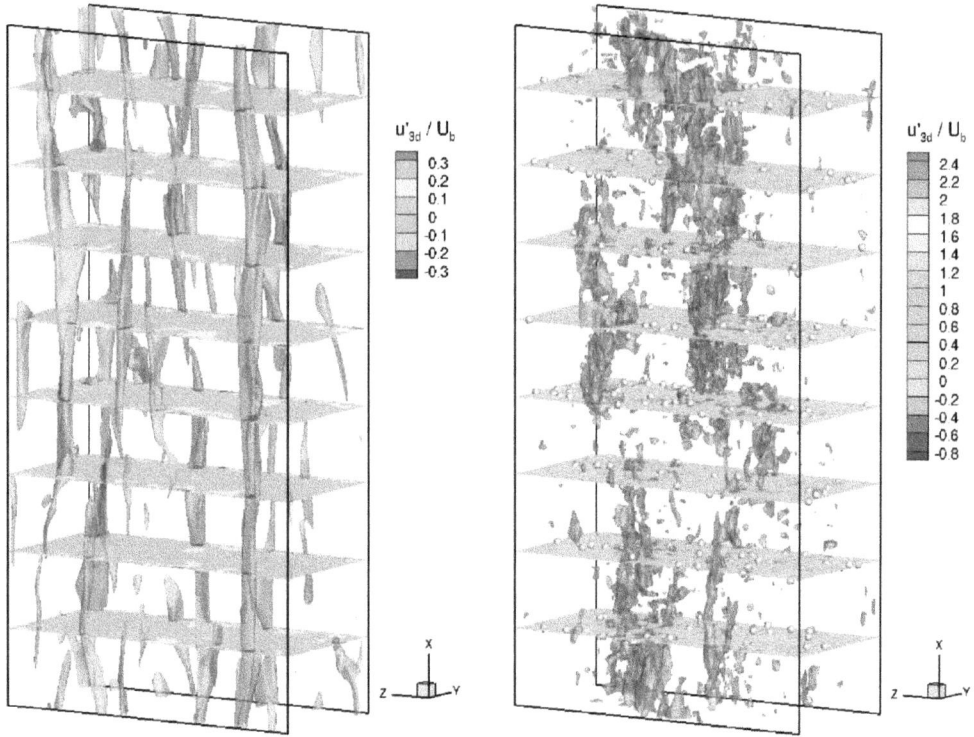

Figure 3.7: Instantaneous three-dimensional velocity fluctuations according to (3.5). Left: *Unladen* case, where the three-dimensional structure is the isosurface $u'_{3d}/U_b = -0.25$. The vertical distance between the planes used for contour plots is $0.5H$. Right: *SmMany*case, with $u'_{3d}/U_b = -0.5$ and different scale for contour plots. Only bubbles cutting the horizontal planes are represented.

influence of bubbles on the large scales is different between the near-wall region and the center region of the present simulation. Near the wall the correlation is always lower in the bubble-laden case since the bubbles rising near the wall disrupt the usual coherent structures. In the center region, the influence of bubbles is more complex. For $0.4 \leq \Delta x/H \leq 1.4$, the correlation is higher when bubbles are present, hinting at some coherent, collective bubble motion. For larger distances $\Delta x/H$, the correlation exhibits a small negative value, suggesting that the presence of bubbles induces elongated structures in the streamwise direction of alternating sign. Uhlmann (2008) found large elongated streamwise structures in simulations of sedimenting particles in channel flow, spanning the whole channel length of $L_x = 4H$ (see Tab. 3.2). In a later study (Garcia-Villalba et al., 2012), the same physics was investigated in a channel of larger streamwise extension, i.e. $L_x = 8H$. It was observed that long elongated structures persist. Most importantly, fluid and particle statistics were not significantly affected by the increase of the channel size. This motivated the choice of the streamwise extent used in the present work.

To obtain further insight into the influence of bubbles on the flow structures, additional averaging of u'_{3d} was performed in the streamwise direction, yielding a two-dimensional velocity

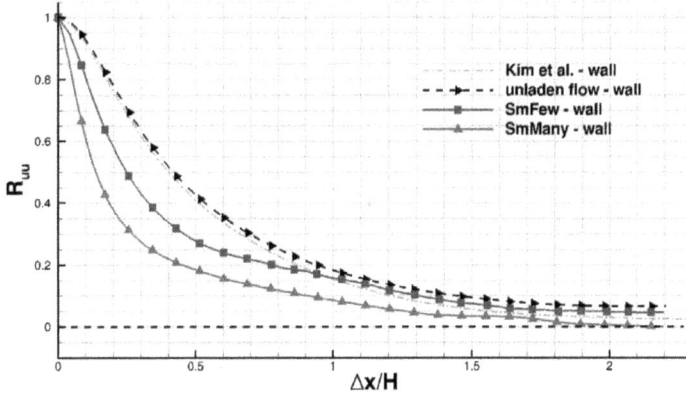

Figure 3.8: Two-point correlation function R_{uu} of the fluctuations of the streamwise velocity in the streamwise direction, according to (3.6). The results are shown for the near-wall region, $y \approx 0.02H$. For the unladen case this means $y^+ \approx 5$. The reference data are from (Kim et al., 1987) with the simulation conducted at slightly different values of the shear Reynolds number, $Re_\tau = 180$ instead of $Re_\tau \approx 168$ in the present case. Results for the case *SmFew* will be discussed in Chap. 4 below.

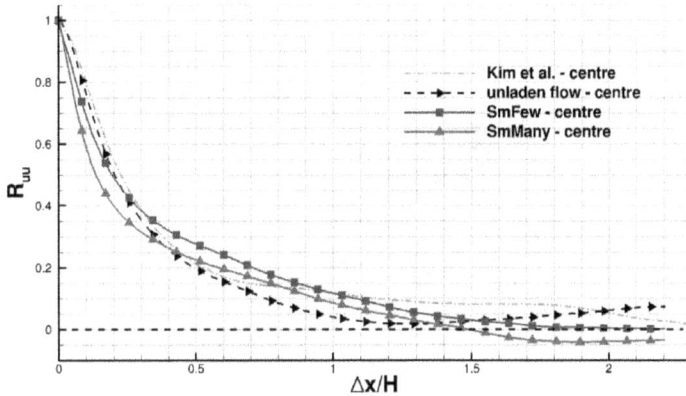

Figure 3.9: Two-point correlation function R_{uu} of the fluctuations of the streamwise velocity in the streamwise direction, according to (3.6). The results are shown for the channel center.

fluctuation

$$u'_{2d}(y, z, t) = \langle u \rangle_x (y, z, t) - \langle u \rangle_{xz} (y, t) . \tag{3.7}$$

This quantity was evaluated in (Uhlmann, 2007) to investigate the life-time of flow structures induced by the particles in a channel flow configuration. Comparing the two-dimensional fluctuations for the unladen and the bubble-laden flow in Fig. 3.10 and 3.11, respectively, the following statements can be made. In the unladen flow, fluctuations are located only in the near-wall regions, with a pattern that is not completely regular, but presents the alternation of regions of negative and positive fluctuations of size approximately 0.2-$0.3H$. For the bubble-laden flow the picture is different. The near-wall structures are replaced by the presence of much larger structures of size $\mathcal{O}(H)$, i.e. covering the entire wall-to-wall gap.

They also alternate in sign in spanwise direction with a period of about $2H$. To quantify the

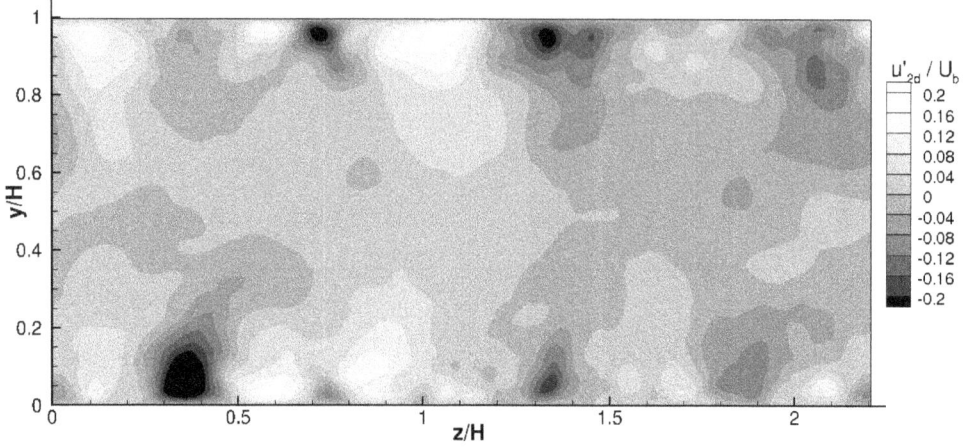

Figure 3.10: Instantaneous two-dimensional velocity fluctuations for simulation *Unladen*, according to (3.7).

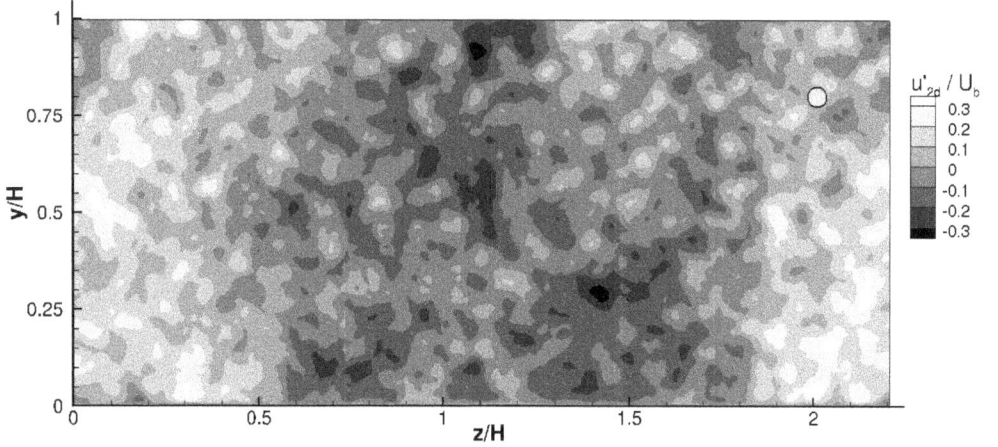

Figure 3.11: Instantaneous two-dimensional velocity fluctuations u'_{2d} according to (3.7) for simulation *SmMany*. The bubble size is represented by a circle at $y/H = 0.8$, $z/H = 2$.

characteristics of the fluid structures shown in Fig. 3.11, the two-point correlation function in spanwise direction $R_{uu}(y, \Delta z)$ was evaluated, defined analogously to $R_{uu}(y, \Delta x)$ in (3.6). As shown in Fig. 3.12, the correlation in the spanwise direction of the unladen flow exhibits a marked minimum around $\Delta z/H \approx 0.15$, i.e. $\Delta z^+ \approx 62$ in very close agreement with the data of Kim et al. (1987), and a return to zero at about twice the distance. With bubbles being present, first of all the decay of R_{uu} in spanwise direction is about twice as fast until $\Delta z/H \approx 0.05$, which is equal to the bubble diameter. Then the decay is almost linear until

$\Delta z/H \approx 0.4$. This result shows that the coherent near-wall structures are disrupted by the interaction between wall turbulence and bubbles. Figure 3.13 shows $R_{uu}(\Delta z)$ in the center

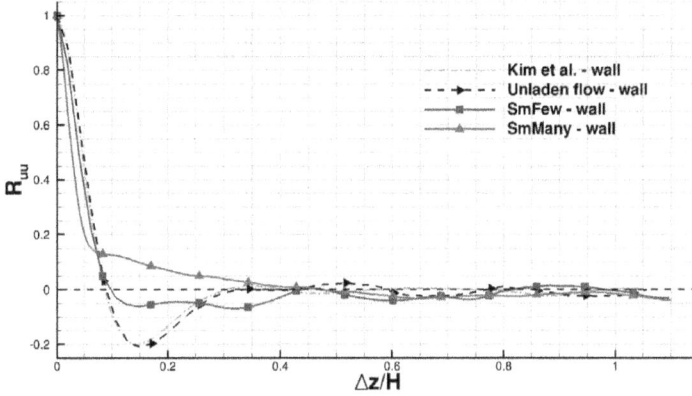

Figure 3.12: Two-point correlation function of the fluctuations of the streamwise velocity in the spanwise direction. The results are shown for the near-wall region, $y \approx 0.02H$. For the unladen case this means $y^+ \approx 5$. The reference data are from (Kim et al., 1987) with the simulation conducted at slightly different values of the shear Reynolds number, $Re_\tau = 180$ instead of $Re_\tau \approx 168$ in the present case.

region. The result for the unladen flow is again validated by the reference data of (Kim et al., 1987), recalling that the slight differences observed are due to the slightly different Reynolds number in both cases, $Re_b \approx 5641$ and $Re_\tau = 180$ instead of $Re_b = 5263$ and $Re_\tau \approx 168$ in the present case. The spanwise correlation for the bubble-laden case exhibits a strong decay until $\Delta z/H = 0.05$ as before, related to the bubble size and hence the size of the wake. For larger distances an almost linear decay is observed between $\Delta z/H \approx 0.1$ and $\Delta z/H \approx 0.8$ with saturation beyond. For $\Delta z/H \geq 0.56$ the correlation is negative and remains so until half the domain size in spanwise direction. So, while near the wall the spanwise coherence is limited to $\Delta z/H \leq 0.4$, truly large structures are observed in the channel center. One period of this phenomenon is captured in the domain providing quantitative support for the observations made with Fig. 3.11. For sedimenting particles with $\phi_{tot} = 0.024\%$, Uhlmann (2007) observed a similar pattern of fluid velocity fluctuations near one wall covering about half the channel width and mirrored near the other wall.

The instantaneous streamwise-averaged bubble concentration $N_{b,2d}$ was evaluated in the yz-plane by averaging the bubble positions over the vertical direction for the instantaneous flow field shown in Fig. 3.7, right, and 3.11. The definition reads

$$N_{b,2d}(y, z, t) = \sum_{i=1}^{N_p} \chi_{[y-I_{2d}/2 \, ; \, y+I_{2d}/2]}(y_p^i(t)) \, \chi_{[z-I_{2d}/2 \, ; \, z+I_{2d}/2]}(z_p^i(t)) , \tag{3.8}$$

where $\chi_{[a;b]}(x)$ is the indicator function assuming the value 1 if $x \in [a; b]$ and 0 else. The quantity I_{2d} is the size of the discretization square employed, set equal to $0.1H$ here. Such visualizations, displayed in Fig. 3.14, and corresponding animations (not reported here) show that regions of high and low two-dimensional fluctuating fluid velocity, as the ones depicted in Fig. 3.11, do not correlate with regions of high and low void fraction.

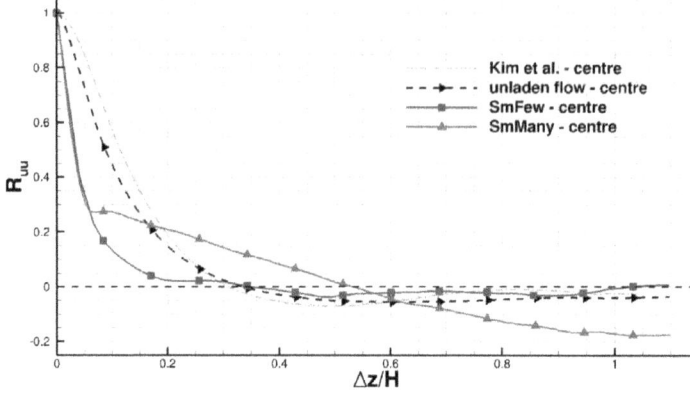

Figure 3.13: Two-point correlation function of the fluctuations of the streamwise velocity in the spanwise direction. The results are shown for the channel center.

A two-dimensional representation of the bubble fluctuations is portrayed in Fig. 3.15, where each bubble position is projected onto the horizontal yz-plane. Bubbles are represented according to the streamwise velocity fluctuation

$$u_p'(x, y, z, t) = u_p(x, y, z, t) - \langle u_p \rangle_{xz}(y) \, . \tag{3.9}$$

It can be observed that regions of positive (negative) fluid fluctuations correspond to region of positive (negative) bubble velocity fluctuations, which means that the instantaneous relative velocity is fairly constant.

3.3.2 One-point statistics of the carrier phase

The instantaneous total wall shear stress was evaluated by means of the streamwise velocity gradient in the wall-normal direction, averaged over both walls,

$$\tilde{\tau}_w = \rho \nu \left\langle \left. \frac{\partial u}{\partial y} \right|_{y=y_w} \right\rangle_{xz} \, . \tag{3.10}$$

Even in the case of bubble-wall collisions, it was always ensured that the position of the streamwise velocity used in (3.10) was outside the bubble. This is due to the collision model described in Sec. 3.1 above, which ensures that the bubble surface is at least one step size of the grid Δ away from the wall, as depicted in Fig. 2.1, right in Sec. 2.3.

The friction velocity and the corresponding value of the shear Reynolds number are obtained from the definitions

$$\tilde{u}_\tau = \sqrt{\frac{\tilde{\tau}_w}{\rho}}, \quad \tilde{Re}_\tau = \frac{\tilde{u}_\tau \, H/2}{\nu} \, , \tag{3.11}$$

with u_τ and Re_τ obtained by additional temporal averaging and collected in Tab. 3.3 for all simulations presented here. The time history of $\tilde{Re}_\tau$ is shown in Fig. 3.16. The initial transient after the introduction of the bubbles in the unladen flow was very short, around $2T_b$. For the case *SmMany* the wall-shear stress is around 23% higher than for the unladen flow, and this is due to the increased level of turbulence generated by the presence of bubbles at

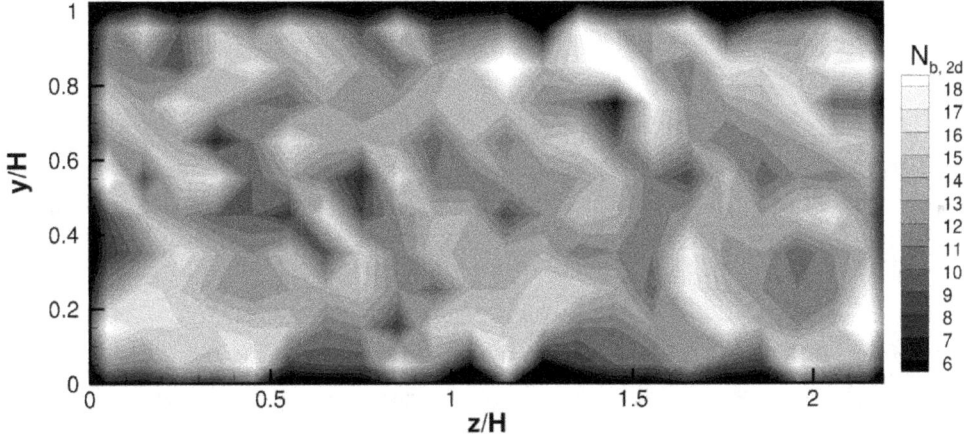

Figure 3.14: Two-dimensional bubble concentration in the yz-plane for the flow field depicted in Fig. 3.11. Discretization width in both directions was set to $I_{2d} = 0.1H$.

the walls, as depicted in Fig. 3.19, left, below. Flow visualizations showed that the highest wall-normal velocity gradients are associated with the wake of bubbles rising near the walls.

Simulation	*Unladen*	*SmMany*	*SmFew*	*LaMany*	*BiDisp*	*SmManyDo*
Re_τ	167.9	209.5	172.3	174.8	194.1	263.9
$RMS(Re_\tau)$	0.34	1.13	0.77	1.54	1.28	1.17

Table 3.3: Time-averaged shear Reynolds number and corresponding RMS-value for all channel flow simulations.

Wall-normal profiles of statistical properties of the fluid velocity field were evaluated and compared with the ones of the unladen channel flow. In the present case, bubbles have almost no influence on the streamwise mean velocity $\langle u \rangle / U_b$, as shown in Fig. 3.17, left. Figure 3.17, right, presents the velocity profile for the unladen and for the bubble-laden flow in a semi-logarithmic plot. These velocity profiles were obtained by averaging over the two sides of the channel and normalizing with the time-averaged values of u_τ. As expected, the shapes are similar but the profile of the bubble-laden flow is shifted below the profile of the unladen flow in this case due to the higher friction velocity. The non-vanishing components of the Reynolds stress tensor are collected in Fig. 3.18. They are higher for the bubble-laden flow, throughout, but the shapes still roughly resemble the ones of the stresses in the unladen case. In the center region, $\langle u'u' \rangle$ is increased by a factor of $34.5 = (5.8)^2$ between *Unladen* and *SmMany*, while the factor is $5.9 = (2.4)^2$ and $7.7 = (2.8)^2$ for $\langle v'v' \rangle$ and $\langle w'w' \rangle$, respectively. It is noteworthy that in the core region $\langle u'v' \rangle$ is not increased by the corresponding factor $5.8 \ast 2.4$. This hints to a much lower correlation coefficient of u' and v' compared to the unladen case. For the streamwise turbulent stress the peaks are at the same distance from the wall. For the other components the maxima of the fluctuations are shifted toward the walls, with the regions of higher fluctuations substantially narrower. The near-wall region with increased fluctuations extends up to a wall distance of $y_w/H \approx 0.15$ and it is most

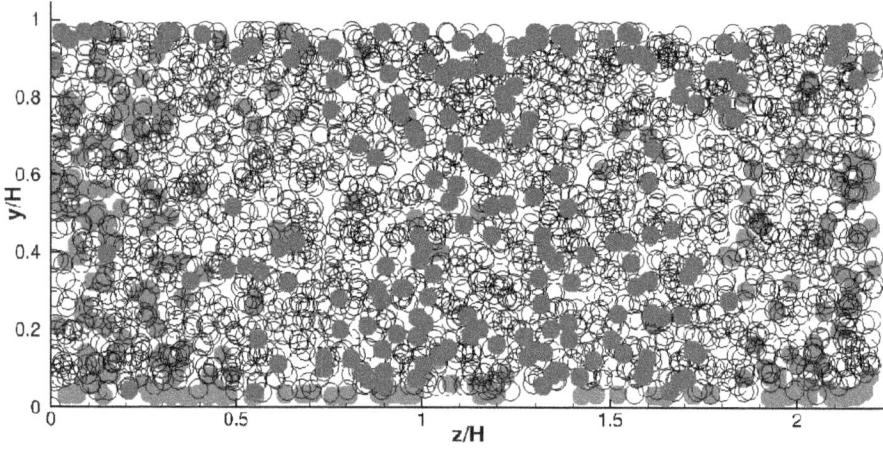

Figure 3.15: Bubble position projected on the horizontal plane and represented differently according to bubble streamwise velocity fluctuations: Blue circle, $u'_p \leq -0.25U_b$; Empty circle, $-0.25U_b < u'_p < 0.25U_b$; Red circle, $u'_p \geq 0.25U_b$. Same flow field as depicted in Fig. 3.11.

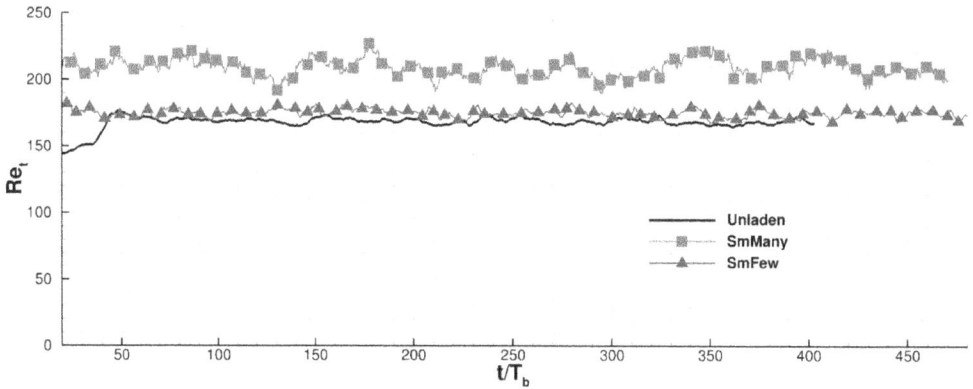

Figure 3.16: Time history of the shear Reynolds number Re_τ according to (3.11).

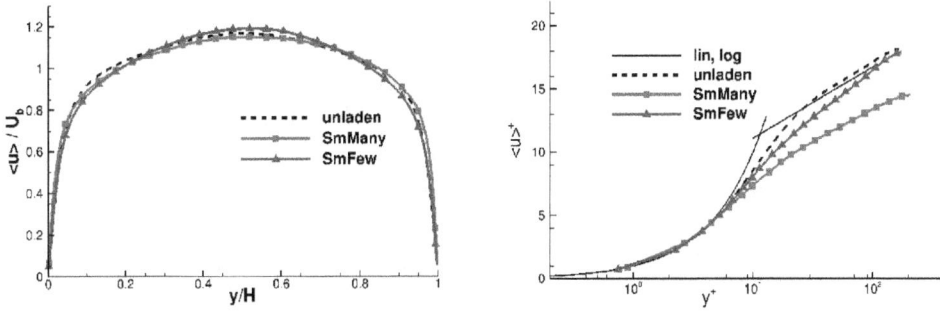

Figure 3.17: Averaged streamwise velocity. Left: Data in bulk units; Right: Data in wall units, averaged over both sides of the channel. The results for the dilute swarm *SmFew* will be discussed in Sec. 4.2 below.

clearly visible with $\langle u'v' \rangle$.

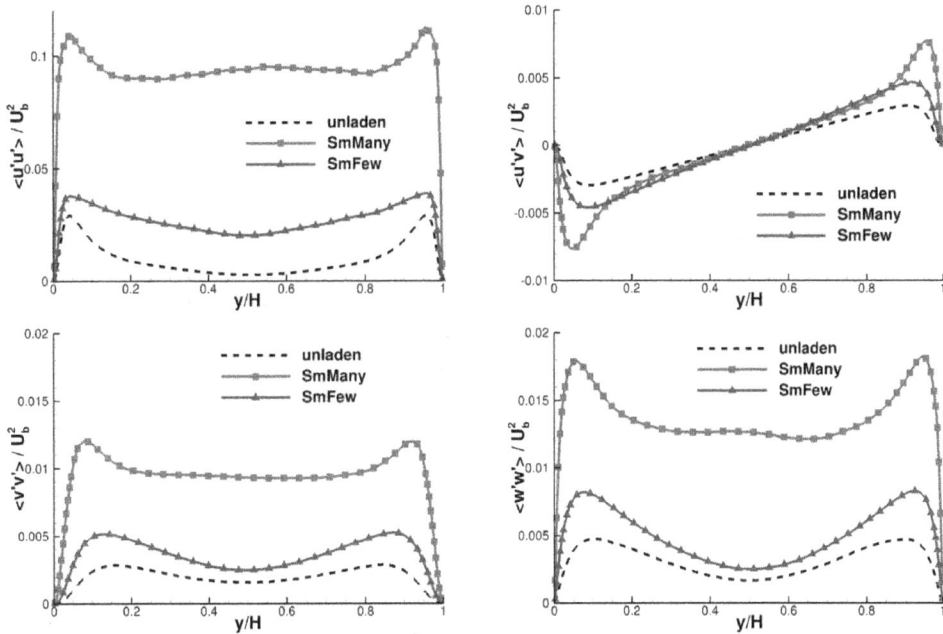

Figure 3.18: Reynolds stress components for *Unladen*, *SmMany* and *SmFew* discussed in Sec. 4.2. Top, left: streamwise fluctuations $\langle u'u' \rangle$; top, right: turbulent shear stress $\langle u'v' \rangle$; bottom, left: wall-normal fluctuations $\langle v'v' \rangle$; bottom, right: spanwise fluctuations $\langle w'w' \rangle$.

The influence of bubbles on the fluid turbulence has been investigated by several authors, reporting both turbulence enhancement and reduction depending on several parameters, like turbulence intensity, turbulent length scales and bubble size, to mention but a few. Hosokawa and Tomiyama (2004) proposed a correlation between the turbulence modification and the ratio R between the eddy viscosity induced by the disperse phase ν_d and the shear-induced eddy viscosity ν_t. In bubbly pipe flows, R can be defined as

$$R = \frac{\nu_d}{\nu_t} = \frac{d_p \, U_r}{u' \, L_\tau} \tag{3.12}$$

where u' is the root mean square of the streamwise fluctuations of the fluid velocity for the single-phase flow, U_r is the bubble relative velocity in the two-phase flow. The definition of U_r according to (3.16) is used here and L_τ is a turbulence length scale. Gore and Crowe (1989) showed that L_τ is about 20% of the pipe radius in turbulent pipe flows, so that $L_\tau = 0.2H/2$ is a reasonable choice here. According to (Hosokawa and Tomiyama, 2004), $R < 1$ implies turbulence reduction, and the breakup of eddies by bubbles is the main mechanism for turbulence modification. For $R > 1$, instead, turbulence enhancement takes place since the turbulence produced by the bubbles is predominant. The ratio R evaluated for the present simulation is $R = 2.73$, when taking the maximum of u' occurring at $y_w/H \approx 0.04$ (Fig. 3.18). As such, R according to (3.12) is a relative crude measure depending on the precise definitions of the quantities involved, so that it should not be over-interpreted. Yet, the fact that turbulence is enhanced in the present case is in good agreement with this approach. Additional information are provide in Chap. 7 below, devoted to the analysis of the TKE modification induced by the bubbles by means of the budget analysis of the TKE transport equation.

3.3.3   One-point statistics of the disperse phase

The following criterion was employed for the frequency at which bubble data were stored: After a stationary state was reached, for a short time a mean bubble velocity was evaluated, by averaging over all bubbles and some laps of time. This mean bubble velocity was used to evaluate how long, on average, a bubble needs to rise a distance equal to its diameter, and this time was used as storage frequency for the bubble data. This yielded a total amount of more than 30000 data sets representing instantaneous bubble fields N_f. Statistical data of the disperse phase were obtained by subdividing the width of the channel into N_w bins and averaging over all bubbles with their center point $\mathbf{x}_p$ located inside the bin, i.e. averaging over the two homogeneous directions and also averaging in time. The mean void fraction hence is determined as

$$\langle \phi \rangle_{xzt} (y) = \phi_{tot} \, \frac{L_y}{N_f \, N_p \, I_w} \sum_{i=1}^{N_p} \sum_{j=1}^{N_f} \chi_{[y-I_w/2 \, ; \, y+I_w/2]}\big(y_p^i(t_j)\big) \, , \tag{3.13}$$

with $y_p^i(t_j)$ the y-coordinate of the center of bubble i at time t_j. Preliminary tests concerning the parameter I_w demonstrated that the results are quite insensitive to this value and after various tests it was set to $I_w = r_p$ to yield high resolution of the computed statistics. Figure 3.19, left, shows that in the simulation *SmMany* the bubbles are distributed fairly evenly with $\langle \phi \rangle \approx \phi_{tot}$ in most of the channel and a peak $\langle \phi \rangle \approx 1.4\phi_{tot}$ near the walls. They hence act similarly as a volume force constant in space (it will be shown below that the relative bubble velocity is almost constant in space) so that the forcing for the case *Unladen* and *SmMany* is very similar. Recall that the volume force driving the flow is adjusted so as to yield the same bulk velocity and hence reduced with the bubbles being present. This is the reason why the mean fluid velocity profiles in Fig. 3.17 of the case *Unladen* and *SmMany* are very similar. The slight asymmetry of the velocity profile is related to the slightly higher

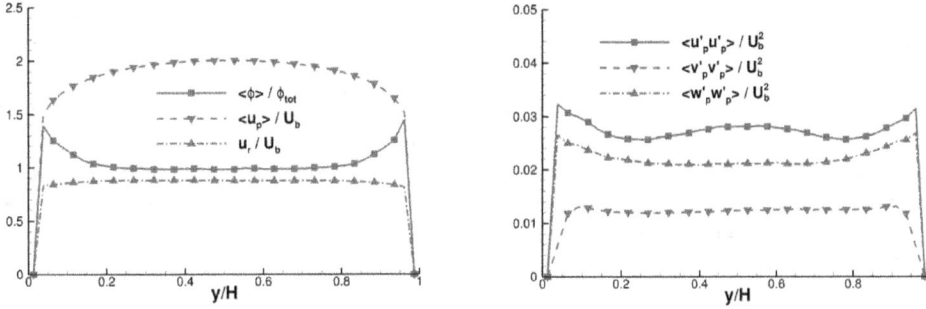

Figure 3.19: Left: Average void fraction distribution, bubble velocity and relative velocity for case *SmMany*. Right: Bubble velocity fluctuations for case *SmMany*. (Symbols represented only for each second data point.)

number of bubbles in the near-wall region for $y \approx H$ compared to $y \approx 0$. This is not a defect of the simulation but related to the bistable nature of such flows, where bubbles can concentrate unevenly over the channel, as commonly observed in bubble column facilities (Bröder and Sommerfeld, 2002). Thought the slight asymmetry, in the near-wall regions the velocity in the *SmMany* case is slightly higher than for the *Unladen* case. This is due to the higher concentrations of the bubbles in these regions, which yield an higher fluid velocity, as observed also in (Dabiri and Tryggvason, 2015). The increased void fraction for wall distances $y_w/H \lesssim 0.15$, suggests that the narrow regions of increased turbulent shear stress $\langle u'v' \rangle$ beyond the linear slope in this near-wall region (Fig. 3.18) are due to the increased forcing introduced by the bubbles. Closely related is the increase of $\langle u'u' \rangle$, $\langle v'v' \rangle$ and also $\langle w'w' \rangle$ in this region. The higher concentration of bubbles near the wall is also the reason why the reduction of the two-point correlation function, shown in Figs. 3.8 and 3.9, is higher in the wall region than in the channel center.

The mean value of any general bubble-related quantity A_p was evaluated in a similar way as the void fraction in (3.13) defining

$$\langle A_p \rangle_{xzt} (y) = \frac{L_y}{N_f N_p I_w} \sum_{i=1}^{N_p} \sum_{j=1}^{N_f} A_p(y_p^i(t_j)) \, \chi_{[y-I_w/2 \,;\, y+I_w/2]}(y_p^i(t_j)) \,. \tag{3.14}$$

A quantity of particular interest in two-phase flows is the relative velocity which is defined as the difference between the bubble or particle velocity and the fluid velocity. While the definition of the particle velocity is obvious, the definition of the fluid velocity at the location of the bubble is delicate for particles not treated as point particles, and several methods have been proposed to overcome this problem. In the case of a fixed object in uniform shear flow, the fluid velocity is usually defined as the upstream velocity on the streamline impinging at the particle center (Kurose and Komori, 1999). For a particle with a diameter of up to 10 times the Kolmogorov length, Bagchi and Balachandar (2004) evaluated the instantaneous relative velocity between the particle velocity and the undisturbed ambient flow. This quantity was taken at the same position in an additional simulation without the particle. A similar strategy was employed by Zeng et al. (2008) investigating the influence of wall turbulence on the forces of a fixed particle in a channel flow. Such an assumption, however, is only justified if the particle has a small influence on the fluid velocity, which is not the case with the simulations presented here. For settling particles in an ambient fluid, Doychev and Uhlmann (2013) defined a time-dependent relative velocity as the difference between the mean particle velocity, averaged over all particles, and the mean fluid velocity, averaged over the domain occupied by the fluid. It appears to be an adequate choice when all Cartesian directions are homogenous, e.g. with periodic boundary conditions, but delicate when the flow is inhomogeneous in at least one direction. Kidanemariam et al. (2013), investigating a horizontal open channel flow laden with heavy particles transported in bed-load mode, defined the characteristic fluid velocity for each particle as the fluid velocity averaged over a so-called "spherical segment surface", defined between two wall-parallel planes at distance equal the bubble diameter. Averaging upstream and downstream velocities, however, does not seem to be adequate when, as in the case of rising bubbles, the two velocities are substantially different, since the downstream flow is constituted mainly of the wake of the bubble. In the present study, we therefore define the relative velocity in an average manner, namely as the difference between the y-dependent mean bubble velocity and the y-dependent mean fluid velocity

$$u_r(y) = \langle u_p \rangle_{xzt}(y) - \langle u \rangle_{xzt}(y) \ . \tag{3.15}$$

To define a characteristic bubble Reynolds number, $u_r(y)$ is averaged over the channel width, but also weighted by the local void fraction $\phi(y)$ to account for the different number of bubbles traveling at a certain relative velocity

$$U_r = \frac{\int_{y=0}^{y=H} u_r(y) \langle \phi \rangle (y) \, dy}{\int_{y=0}^{y=H} \langle \phi \rangle (y) \, dy} \ , \tag{3.16}$$

with

$$Re_p = \frac{U_r \, d_p}{\nu} \ . \tag{3.17}$$

The results obtained for the different cases are collected in Tab. 3.2. Figure 3.19, left, presents the bubble velocity and the relative velocity for the simulation *SmMany*. As expected, the rise velocity of the bubbles in the channel center is higher due to the higher fluid velocity. Nevertheless, the relative velocity is almost constant over the channel and only slightly higher at the channel center.

Simulation	*SmMany*	*SmFew*	*LaMany*	*BiDisp* SM	*BiDisp* LA	*SmManyDo*
Re_p	235.5	268.3	475.2	233.6	463.6	221.3

Table 3.4: Bubble Reynolds number Re_p according to (3.17).

Bubble velocity fluctuations in the streamwise direction, u'_p, are defined according to

$$u'_p(y) = u_p(y) - \langle u_p \rangle_{xzt}(y) \ . \tag{3.18}$$

with analogous definitions for the velocity components v_p and w_p in the y- and in the z-directions, respectively. This yields the velocity variances collected in Fig. 3.19, right. The fluctuations in streamwise direction are higher but comparable to the ones in the spanwise direction. This is in pronounced contrast to the fluid velocity fluctuations (Fig. 3.18) where $\langle u'u' \rangle$ is by a factor of about 10 larger then the other components. When bubble and velocity fluctuations are compared, in the channel center $\langle u'_p u'_p \rangle \approx 10 \langle u'u' \rangle$, while $\langle v'_p v'_p \rangle \approx \langle v'v' \rangle$ and $\langle w'_p w'_p \rangle \approx 1.7 \langle w'w' \rangle$.

3.3.4 Mechanism determining the void fraction distribution

Several mechanisms contribute to the spatial distribution of particles and bubbles in a fluid as reviewed in (Guha, 2008). The most important ones are the lift force generated by a gradient of the fluid velocity, the turbophoresis effect created by a gradient of mean fluctuations, the turbulent diffusion generated by a gradient of the concentration of the disperse phase, even with spatially constant root mean square values of the fluid velocity, and collision effects. To understand the wall-normal distribution of bubbles in the present turbulent channel flow, the shear-induced lift force, the influence of bubble-wall and bubble-bubble collisions and the turbophoresis effect are discussed now.

Shear-induced lift Tomiyama et al. (2002) experimentally investigated the lateral migration of bubbles in simple shear flow in a tank between a fixed wall and a moving belt. They discovered that small, spherical bubbles with diameter smaller than $5mm$ moved toward the fixed wall, while larger bubbles, with an ellipsoidal-like shape, moved toward the moving wall, i.e. toward the region of higher fluid velocity. This feature is related to the deformation, i.e. to the bubble shape, as investigated in (Adoua et al., 2009). Numerical simulations (Tanaka, 2011; Lu and Tryggvason, 2013) confirm this tendency in channel flow configurations. The knowledge acquired by such studies can hardly be applied to the present case, however, since important differences have to be considered. The bubbles investigated in the present study have a constant spherical shape, and at the phase boundary a no-slip condition is imposed, different from the free-slip conditions applied in the cited references. Another difference is the bubble Reynolds number, which is much higher in the present simulations. To clarify this issue a separate simulation was performed to investigate the response of a fixed bubble in a cross flow with constant shear rate at the same Reynolds number as for the bubbles in the channel, as reported in Sec. 3.2. It was shown that the lift force is directed towards the side of the bubble with the lower relative velocity for the regime investigated here. As described in Sec. 3.2, in a channel flow the lift force is supposed to be directed toward the core region that is the side of the bubble that, on average, experiences the lower relative velocity. The void fraction maxima, however, are close to the walls which hence can not be explained in this way.

Collisions The influence of collisions on the void fraction distribution in the present simulation was investigated by evaluating the number of the collision events and their contribution. Indeed, the number of collision events is extremely small. The average number of bubbles colliding with the walls for one instantaneous flow field is about 7.2 on average, which is quite low observing that typically around 220 bubbles are located in each of the two bins closest to the walls where bubble-wall collisions can happen. The average number of bubble-bubble collision events is only about 2.2 in the entire flow field, which is negligible considering the total number of 2880 bubbles in the flow. The mean wall-normal distribution of the collision events (not shown here) resembles the one of the void fraction. These data confirm that collision events are so rare that they only play a very marginal role in the flow, if any.

Turbophoresis Young and Leeming (1997), in the context of particle deposition in turbulent pipe flows, derived the time-averaged particle transport equation in the radial direction

$$\langle v_r \rangle \frac{\partial \langle v_r \rangle}{\partial r} = -\frac{\langle \Phi_D \rangle \langle v_r \rangle}{\tau_p} + \langle F_r \rangle - \frac{\partial}{\partial r} \langle v_r' \, v_r' \rangle + \frac{\langle v_\phi' \, v_\phi' - v_r' \, v_r' \rangle}{r} - \langle v_r' \, \mathbf{V} \cdot \nabla(\ln \rho_p) \rangle \ , \quad (3.19)$$

where v_r is the particle velocity in the radial direction and r is the radial coordinate. The first term on the right hand side of (3.19) is related to the drag force, the second accounts for the lift force, the third represents the turbophoresis effect, the fouth is associated to the fluctuations of the particle velocity in the radial and in the azimuthal direction (v'_ϕ), and the fifth is related to the inomogenities of the particle density ρ_p (the product of particle number and particle mass) and to the mean particle velocity $\mathbf{V}$. Equation (3.19) reflects the mechanisms involved in the distribution of particles and here we focus on the so-called turbophoresis term. As stated in (Young and Leeming, 1997), it represents the tendency of particles to acquire a drift velocity in the direction of decreasing turbulence intensity. An analysis of this term was motivated by the observation that the spatial distribution of dispersed objects is influenced by a gradient of the fluctuating flow: Particles tend to move toward regions where the turbulence is lower, since it is more probable for them to obtain a higher momentum in a region with pronounced turbulence than in a low-turbulence region. For the present channel flow configuration the radial direction corresponds to the wall-normal direction and the turbophoresis term can be evaluated as

$$T(v'_p) = -\frac{\partial \langle v'_p \, v'_p \rangle}{\partial y} \ . \tag{3.20}$$

To illustrate the influence of $T(v'_p)$ only half of the channel is considered, $0 < y/H < 0.5$, and a central second-order approximation was employed to evaluate the derivative. If $T > 0$, particles are pushed toward increasing y, i.e. toward the center region. If $T < 0$, they are pushed toward decreasing y, i.e. toward the wall. Figure 3.20 presents the wall-normal distribution of T and $\langle \phi \rangle$. It can be recognized that the profile of T has three regions:

- For $0 < y/H \lesssim 0.09$, T is negative moving the bubbles toward the wall.

- For $0.09 \lesssim y/H \lesssim 0.26$, T is positive moving the bubbles toward the center region, but at a much lower rate, since the absolute value of T is smaller.

- For $0.26 \lesssim y/H < 0.5$, T is almost zero, having no influence on the bubble distribution.

According to Fig. 3.20, the influence of the turbophoresis is higher in the near-wall region, where T assumes large negative values. The fact that the gradient of $\langle \phi \rangle / \phi_{tot}$ is still negative results from the other terms in (3.19) which have an equilibrating effect.

To further support the discussion of the turbophoresis effect this quantity was also evaluated with the wall-normal derivative of the fluid wall-normal fluctuations as

$$T(v') = -\frac{\partial \langle v' \, v' \rangle}{\partial y} \ . \tag{3.21}$$

This evaluation is preferred when the mean-square fluctuation velocity of the particle is not known (Young and Leeming, 1997) and is employed, for example, by Nowbahar et al. (2013) to address the role of the turbophoresis on the distribution of small inertial particles in channel flows. The profile of T according to (3.21) is also reported in Fig. 3.20. Both profiles of the turbophoresis term present the same attitude according to which three different zones are observed and the crossings of the zero-line of the profiles are at the same position, backing up the assumption of the role of the turbophoresis effect on the bubble distribution. This detailed discussion demonstrates that the gradient of the fluctuations of the velocity of the bubble and of the fluid moves the bubbles toward the wall for $y/H \lesssim 0.09$ (and symmetrically for $y/H \gtrsim 0.91$). It is hence this effect which is responsible for the increased void fraction distribution near the walls in Fig. 3.19.

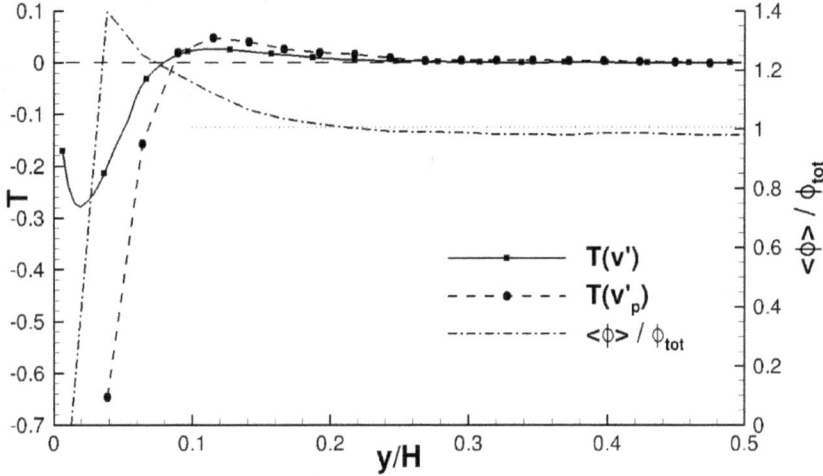

Figure 3.20: Quantification of the turbophoresis effect for simulation *SmMany*. Wall-normal profile of T according to (3.20) and (3.21) evaluated with a central difference approximation and void fraction. Dashed line: $T = 0$. Dotted line: $\langle \phi \rangle = \phi_{tot}$.

3.3.5 Pair correlation functions

When the flow of disperse objects is investigated, may they be particles or bubbles, one of the most important aspects is the spatial distribution of the disperse phase and how this is related to the characteristic of the surrounding fluid. Several approaches have been developed for the analysis of preferential concentration and bubble clustering and a detailed review was provide by Monchaux et al. (2012) and recently Santarelli et al. (2014a) proposed a method for the detection of bubble clusters based on a smoothed void fraction approach. In this section, the mutual position and the interaction of bubbles is investigated by variants of the pair correlation function (PCF) which is a common tool for the analysis of bubble pairs, as in (Bunner and Tryggvason, 2003; Martinez Mercado et al., 2010). The radial-PCF (r-PCF) is a measure of the probability to find a bubble pair at a certain distance. The definition reads:

$$G_r(r) = \frac{\Omega}{N_p(N_p - 1)} \frac{1}{\Delta V(r)} \sum_{i=1}^{N_p} \sum_{\substack{j=1 \\ j \neq i}}^{N_p} \chi_{[r-\frac{1}{2}\Delta r; \, r+\frac{1}{2}\Delta r]}(d_{i,j}), \qquad (3.22)$$

where r is the radial variable, Ω is the volume of the investigated sphere, Δr the step size in r, $\Delta V(r) = (4\pi/3)[(r + \Delta r/2)^3 - (r - \Delta r/2)^3]$ the volume of the spherical shell of inner radius $r - \Delta r/2$ and outer radius $r + \Delta r/2$, and $d_{i,j}$ the distance between the center of bubble i and the center of bubble j. For infinitely small randomly distributed bubbles in an unbounded domain, the r-PCF takes the value 1 for all distances r. In the present case, two differences arise when evaluating the r-PCF. The first, is that bubbles have a volume so that they can not overlap. This yields $G_r(r) = 0$ for $d_{i,j} \leq 2$. The second is related to the presence of the walls. Even for randomly distributed bubbles inside the channel, the lack of bubbles beyond the wall yields a value lower than 1 if the distance from the wall is smaller than the investigation size, i.e. if $r > \min\{y_p, H - y_p\}$ in (3.22).

Figure 3.21 shows the r-PCF for the bubbles in the present channel flow compared to randomly distributed bubbles (RDB), i.e. to a random distribution of bubbles in the channel geometry, under the restriction that $d_{i,j} > 2$. The analysis is limited to $r < H/2$ for obvious reasons, so that $\Omega = 4\pi(H/2)^3/3$ in (3.22). Note that with (3.22) averaging is performed over all bubbles in the flow field to yield a single curve characterizing the behavior. The most probable pair distance (MPPD) is $4r_p$. When compared to the randomly distributed bubbles, three zones can be detected. In the first zone, $2r_p \lesssim r \lesssim 3r_p$, the r-PCF is smaller than the random distribution: Small-scale mechanisms increase the distance between two bubbles. For $3r_p \lesssim r \lesssim 15r_p$ bubble pairs are more probable than for the random distribution, suggesting this length scale to be the clustering length scale for the present channel flow configuration. For larger distances, the PCF is slightly below the random case so that no cluster of the size is expected.

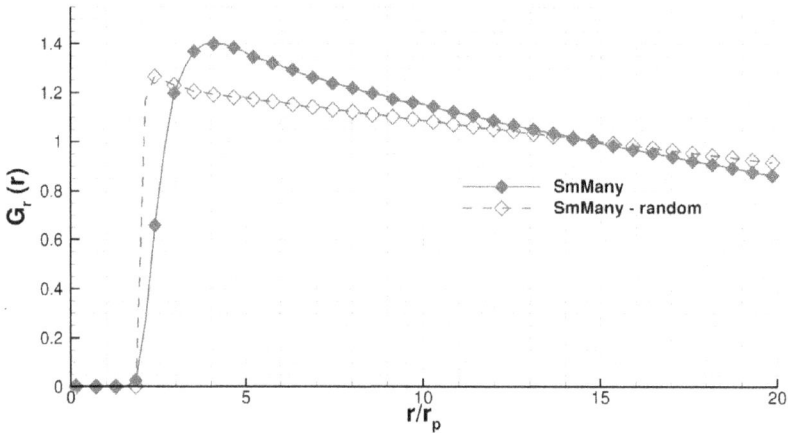

Figure 3.21: Radial PCF computed for case *SmMany* and RDB, i.e. random distribution of bubbles in the channel.

The small-scale repulsion and the attraction of bubbles on intermediate scales can be explained by the analysis of the flow field around two fixed spheres in cross flow. As described by Kim et al. (1993), two spheres repel each other when the distance between the surfaces is small. This is due to the different fluid acceleration on the two sides of the sphere, i.e. a lower acceleration in the region between the bubbles and a higher acceleration in the external region. It is caused by the position of the stagnation point on each sphere, which moves toward the neighboring bubble. When the distance between the bubbles increases, the picture is different. The position of the stagnation point is not influenced by the presence of the other bubble any more and the acceleration effect is negligible. In this case, the velocity in the region between the two spheres is higher with respect to the velocity in the external region, leading to a low-pressure zone between the spheres. Hence, this pressure-related mechanism, also called Bernoulli effect, leads to the attraction of the spheres. The distance at which the flow behavior changes, i.e. at which no force is experienced by the spheres, depends on the Reynolds number of the sphere and it decreases with increasing Re_p. For the largest Reynolds number investigated in (Kim et al., 1993), $Re_p = 150$, bubbles attract each other for $d_{i,j}/r_p > 6.8$. It can therefore be assumed that this distance is even smaller for higher Re_p, thus explaining the distance $r \lesssim 3r_p$ found in the present configuration, where bubble pairs are less frequent in Fig. 3.21.

Additional information on the interaction of bubbles can be obtained by the evaluation of the angular PCF (a-PCF) which measures the probability of the angle between two bubbles. According to the usual definition of the a-PCF in (Bunner and Tryggvason, 2002a), the probability of the pair angle is evaluated for all bubble pairs whose distance is below a certain distance. As proposed by Santarelli and Fröhlich (2013), the definition of the a-PCF is modified here to account only for bubble pairs within a certain spherical shell. The definition then reads

$$
\begin{aligned}
G_\phi(\phi_g; r_1, r_2) ={}& \frac{\Omega_s}{N_p(N_p - 1)} \frac{1}{\Delta V(\phi_g, r_1, r_2)} \cdot \\
& \cdot \sum_{i=1}^{N_p} \sum_{\substack{j=1 \\ j \neq i}}^{N_p} \chi_{\left[\phi_g - \frac{1}{2}\Delta\phi_g; \, \phi_g + \frac{1}{2}\Delta\phi_g\right]}(\Phi_{i,j}) \, \chi_{[r_1; \, r_2]}(d_{i,j})
\end{aligned}
\tag{3.23}
$$

where r_1 and r_2 are the radial distances defining the investigation shell, Ω_s the volume of the shell between the sphere of radius r_1 and r_2. Furthermore, ϕ_g is the angular variable, $\Phi_{i,j}$ the angle between the line connecting the center of bubble i and the center of bubble j with respect to the vertical direction, $\Delta\phi_g$ the step size in ϕ_g, and $\Delta V(\phi_g, r_1, r_2) = 2/3\pi[(r_2)^3 - (r_1)^3] \, [\cos(\phi_g - \frac{1}{2}\Delta\phi_g) - \cos(\phi_g + \frac{1}{2}\Delta\phi_g)]$. The origin of the angular coordinate ϕ_g is the positive vertical direction, yielding $\phi_g = 0$ and $\phi_g = \pi$ for vertically aligned bubbles and $\phi_g = \pi/2$ for horizontally aligned bubbles. The a-PCF according to (3.23) takes the value 1 for each angle in case of infinitely small bubbles randomly distributed in an unbounded domain. The a-PCF of randomly distributed bubbles in the present domain (not shown here) is only marginally influenced by the boundedness of the domain, yielding an almost constant value, slightly higher than 1 for vertical alignment and slightly lower for horizontal alignment, as expected when considering the influence of the vertical walls.

Besides the canonical representation of G_ϕ as a function of the angle ϕ for a given investigation region, shown in Fig.3.22, left, a two-dimensional illustration of the a-PCF is provided in Fig. 3.22, right, where the two coordinates are the distance $\rho_r = (r_2 - r_1)/2$, with $r_2 - r_1 = r_p$ chosen here, and the angle ϕ_g. This representation allows a more intuitive understanding of the frequency of pair alignments.

For the canonical representation (Fig. 3.22, left) three radial shells were investigated. The first one, $0 < d_{i,j}/r_p \leq 3$, accounts for small-scale interactions of bubbles. When the distance is small, bubbles tend to align mainly horizontally and, additionally, the a-PCF presents two local maxima for $\phi_g = 0$ and $\phi_g = 180°$, i.e. for vertical alignment. In the two-dimensional graph in Fig. 3.22, right, the preferential horizontal alignment is indicated by the high values of the a-PCF for small ρ_r and $\phi_g = 90°$. The preferential horizontal alignment is a common feature for spherical rising bubbles, as described in (Esmaeeli and Tryggvason, 2005), where simulations of rising bubbles in a triple periodic domain were performed. This feature is caused by the pressure field around a rising pair, as discussed above. At small distances, the a-PCF has two local maxima, for $\phi_g = 0$ and $\phi_g = 180$ and this is due to the so-called "drafting kissing and tumbling" mechanism, which can be also be experienced by bubbles, as described in (Cartellier and Riviere, 2001). Furthermore, the vertical alignment

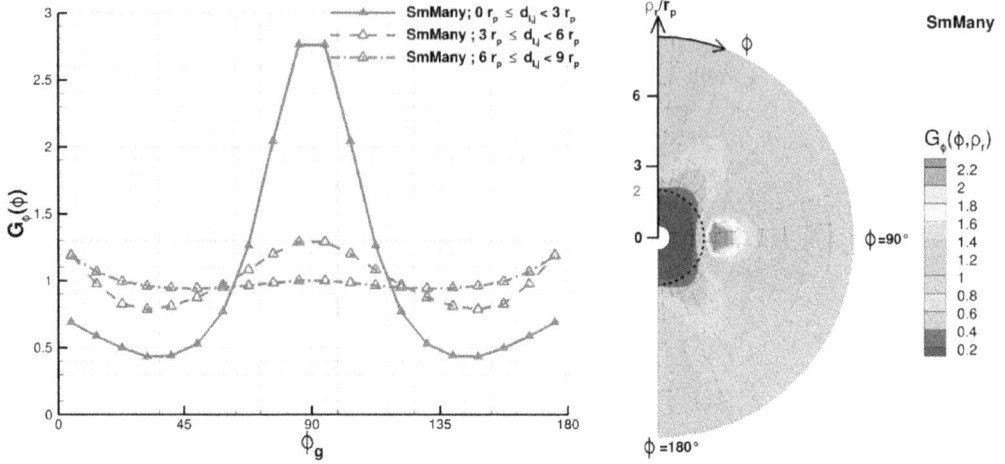

Figure 3.22: Angular PCF for *SmMany*. Left: Canonical representation. Different curves for $d_{i,j}/r_p \in [0;3[$, $d_{i,j}/r_p \in [3;6[$ and $d_{i,j}/r_p \in [6;9[$. For this picture $r_2 - r_1 = 3\,r_p$ in (3.23). Right: Two-dimensional representation, with $\rho_r = (r_2 - r_1)/2$ and $r_2 - r_1 = r_p$.

of bubbles was previously described as an equilibrium configuration, although an unstable one, by Bunner and Tryggvason (2002a). The in-line configuration is related to the suction mechanism that the wake of the leading bubble exerts on the trailing bubble. For increasing distances, both horizontal and vertical alignments are reduced, and for $9 < d_{i,j}/r_p \leq 12$ (not shown here), the a-PCF of the bubbles is indistinguishable from the one of randomly distributed bubbles. This confirms that bubble interaction and potential bubble clustering takes place on small scales, i.e. for distance smaller than $10\,r_p$. The collected results provide quantitative data on this phenomenon suitable as reference for model development and validation.

In conclusion, the influence of the bubbles has been investigated both for the fluid and the disperse phase and the following observations can be drawn. The introduction of a bubble swarm strongly modifies the flow structures usually observed in an unladen configuration. The usual channel flow structures are disrupted by the bubbles, the bubble-induced elongated structures are quite large and, hence, the streamwise correlation of the fluctuating fluid velocity is larger than zero. The turbulence of the fluid phase is enhanced by the presence of the bubbles, since the turbulence generation related to the bubbles is dominant for the investigated parameter range, in accordance with other studies in the literature. The turbophoresis effect, related to the wall-normal gradient of bubble velocity fluctuations, was proved to dominate over the lift force in determining the wall-normal bubble distribution. In the present swarm, bubbles align mainly horizontally due to the pressure-related aspiration mechanism between the bubbles.

4 Influence of void fraction

In the present chapter a second simulation is presented where a dilute swarm of small bubbles is considered. This simulation, labeled *SmFew*, is analyzed and compared with the *SmMany* and *Unladen* case to investigate the influence of the total void fraction on both the fluid and on the bubble dynamics. The total void fraction is reduced by factor 7.5 with respect to the *SmMany* case, as reported in Tab. 3.2, and this allow a detailed investigation of the influence of the void fraction on the flow features. As for Chap. 3, some of the results reported in this chapter have been presented in Santarelli and Fröhlich (2014).

4.1 Fluid phase: Flow structures and correlation functions

Three-dimensional fluid fluctuations evaluated for the dilute swarm *SmFew* according to (3.5) are shown in Fig. 4.1. Two types of structures are present in the flow: Near-wall structures which resemble the ones of the unladen flow and larger structures generated by the bubble presence. A comparison can be made with the picture provided in Fig. 3.7 above for the *SmMany* and for the *Unladen* cases. The structures are smoother than the corresponding ones in the denser swarm and this is related to the fluctuation level which is lower than in the *SmMany* case due to the smaller number of bubbles, as will be further discussed in Sec. 4.2 below. Compared with the unladen channel, longitudinal structures appear, induced by the presence of the bubbles.

The two-point correlation function in the streamwise direction reported in Fig. 3.8 and 3.9 allow the following statements. The curve of the streamwise correlation R_{uu} of *SmFew* is located between the one of the unladen flow and the one of the denser swarm *SmMany*, as expected. The values are somewhat closer to the ones of *SmMany* for the short range, up to about $\Delta x / H = 0.7$. This confirms that the small-scale reduction is due to the turbulence generated by the bubbles, the latter increasing with the global void fraction. The large-scale behavior resembles the one of the unladen flow and the correlation function presents a small, positive value. In the center region, the picture is not so clear. For small scales, $0 < \Delta x / H < 0.2$, the reduction due to bubble-generated turbulence is confirmed. But for distances $0.2H < \Delta x < 1.5H$ the correlation is higher than for both, the *Unladen* and the *SmMany* simulation. Nevertheless, for large distances, the correlation is almost zero and this indicates that in the dilute swarm the bubbles induce flow structures smaller than the ones induced by the denser swarm.

Two-dimensional fluctuations according to (3.7) are presented in Fig. 4.2. The graph shows near-wall structures similar to the unladen flow (Fig. 3.10) albeit somewhat larger. While no activity is seen in the channel center for the unladen flow, this is substantially different

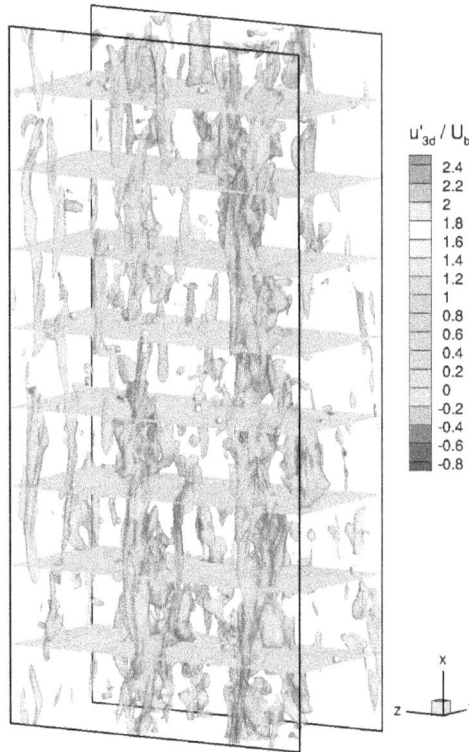

Figure 4.1: Isosurfaces of instantaneous three-dimensional velocity fluctuations defined by (3.5) for simulation *SmFew* with $u'_{3d}/U_b = -0.25$. The horizontal planes inserted at distances of $0.5H$ show contour plots of u'_{3d}/U_b. Only bubbles cutting such planes are represented.

here, with $\phi_{tot} = 0.29\%$. The pattern of longitudinal coherence exhibits smaller scales in the horizontal plane as for *SmMany*. This impression is confirmed by the two-point correlation function in the spanwise direction, depicted in Fig. 3.12 and 3.13. In the near-wall region it attains values between those of the cases *Unladen* and *SmMany* up to $\Delta z/H = 0.25$. For larger distances the behavior is somewhat different showing further undulations of the curve. It was taken care that these do not result from insufficient averaging. The data of *SmFew*, hence, suggest a combination effect between the narrower streaks of the unladen flow and the wider structures found in the denser swarm. The correlation function in the center region, portrayed in Fig. 3.13, is identical to the one of the denser swarm for $z/H < 0.06$ and this means that for very short distances the reduction of the correlation with respect to the single-phase flow is not affected by the number of bubbles. For larger distances, the correlation function for the dilute swarm equals the one of the single-phase flow with very small negative values. Hence, in the present case, no large-scale feature is observed. As for the denser swarms, no apparent correlation was found between fluctuation regions and two-dimensional bubble distribution, $N_{b,2d}$, for an instantaneous flow field (picture not reproduced).

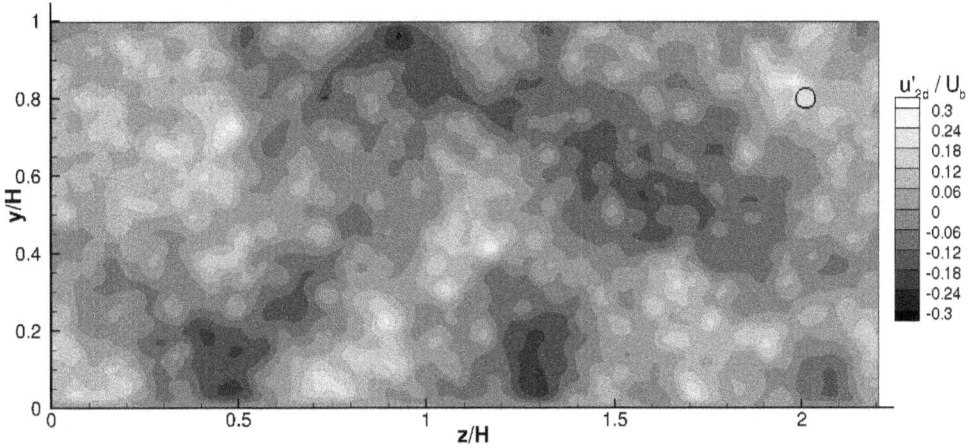

Figure 4.2: Instantaneous two-dimensional velocity fluctuations u'_{2d} according to (3.7) for simulation *SmFew*. The bubble size is represented by a circle at $y/H = 0.8$, $z/H = 2$.

4.2 One-point statistics of the carrier phase

With $Re_\tau = 172.3$, the wall shear stress for *SmFew*, is around 3% higher than in the unladen case (Tab. 3.2). The mean streamwise velocity profile, portrayed in Fig. 3.17, is slightly modified with respect to the unladen flow. The velocity profile in wall units is presented in Fig. 3.17, right. As for the other case, a logarithmic region occurs for $y^+ > 20$. As a consequence of the higher fluid velocity in the center, the slope of the logarithmic range is substantially higher than for the two other cases. The non-vanishing components of the Reynolds stress tensor are presented in Fig. 3.18. The shape of all profiles resembles the one of the unladen flow, but all components are enhanced. The enhancement is due to the turbulence produced by the bubbles, as described in Sec. 3.3.2. The fluctuations are smaller than in the *SmMany* simulation, and for this regime and configuration the turbulence enhancement due to the bubbles increases with the void fraction without changing the shape of the profile too much. The maxima in all curves are less pronounced than for the *SmMany* and the wall-normal coordinates of their occurrence are between the respective values of *Unladen* and *SmMany*. The maxima for the streamwise component of the Reynolds stress are at the same wall-normal position as in the unladen flow. In the core region the components $\langle v'v' \rangle$ and $\langle w'w' \rangle$ present the same value, as for the unladen channel, suggesting an isotropic behavior of the fluid turbulence in this region. This feature was not observed for the denser swarm where, in the core region, fluid fluctuations were larger in the spanwise direction.

4.3 One-point statistics of the disperse phase

The void fraction distribution, the rise velocity and the relative velocity of the bubbles in the dilute swarm are presented in Fig. 4.3, left. The bubbles have an almost uniform distribution for $0.1 < y/H < 0.9$, and within this region, a slightly higher concentration is present around $y_w/H \approx 0.19$. Very close to the wall, $\langle \phi \rangle$ is substantially smaller than in the

SmMany case. As expected, the relative velocity u_r is almost uniform but very slightly higher at the wall regions, in contrast to the *SmMany* case, where it slightly decays near the walls. A substantial absolute bubble velocity of $\langle u_p \rangle \approx 1.7 U_b$ is observed near the walls, increasing up to $2.2 U_b$ in the channel center. When comparing the relative velocity for *SmFew* and

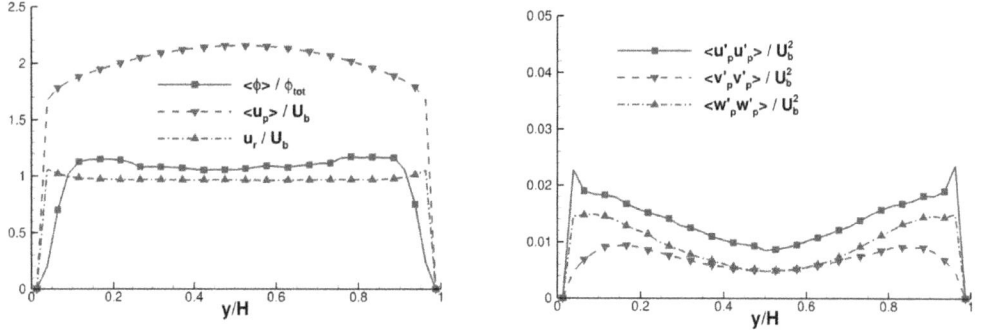

Figure 4.3: Left: Average void fraction distribution, bubble velocity and relative velocity for case *SmFew*. Right: Bubble velocity fluctuations for *SmFew* case. (Symbols represented only for each second data point.)

SmMany, it can be observed that the swarm with the smaller number of bubbles rises, on average, faster than the denser swarm. This is due to the hindrance effect: The higher the number of bubbles, the lower the mutual distances and the larger are the fluid shear stresses at the bubbles surfaces. Ishii and Zuber (1979) proposed a quantification of the effect for sedimenting particles as a function of the total void fraction by means of a power law in the form of $(1 - \phi_{tot})^\alpha$, where the value of α depends on the flow regimes. Garnier et al. (2002) used the quantity

$$\Delta E = \frac{Re^* - Re_p}{Re^*} \qquad (4.1)$$

to measure this effect, here formulated in terms of the Reynolds number. In the present work, Re^* is the Reynolds number of a bubble rising in infinite liquid evaluated with the terminal velocity and set equal to the Reynolds number of a bubble rising in a quiescent fluid under otherwise the same conditions. The value Re^*, hence, was set equal to $Re_{p,\infty}$ evaluated in Sec. 3.2.1 for the simulation *SingleSm* while Re_p is the bubble Reynolds number according to (3.17). Table 4.1 collects the results for *SmFew*, for *SmMany* and for a third simulation with $\phi_{tot} = 1\%$ and the same bubble size as *SmFew* and *SmMany* which will not be discussed further here.

Simulation	*Single*	*SmFew*	*SmMed*	*SmMany*
ϕ_{tot}		0.29%	1%	2.14%
Re_p	265.7	268.3	257.6	235.5
ΔE	-	-0.005	0.032	0.102

Table 4.1: Quantification of the hindrance effect according to ΔE from (4.1). The simulation labeled *SmMed* was performed in preliminary work under the same conditions as the simulations *SmFew* and *SmMany*, except for the domain extensions in the x- and z-direction, being smaller by a factor 4 with respect to the channel extensions reported in Tab. 3.1.

Figure 4.4 provides a comparison of the present data with the experimental work of Garnier

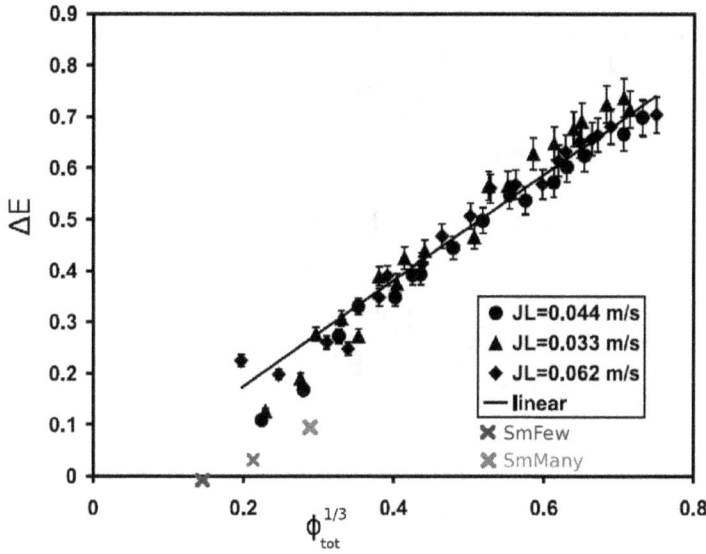

Figure 4.4: Quantification of the hindrance effect according to (4.1): comparison between numerical results and experiments from Garnier et al. (2002). [Usage of Figure with permission from Elsevier.]

et al. (2002) and the results exhibit good qualitative agreement. The observed qualitative differences con be due to several issues: Pipe flow in the experiments and channel flow in the simulation, the higher bubble Reynolds number in the experiments, the different evaluation of the relative velocity, etc. The reduction of the swarm rise velocity in the dilute case is very small due to the very low void fraction: Mutual bubble distances are relative large and the pressure-related mechanism, discussed in Sec. 3.3.5, is not the dominant effect, since it is a short-range effect. The wake-aspiration meachanism, instead, dominates here since the influence of a bubble wake is felt by another bubble to larger distances, as portrayed in Fig. 4.5. This phenomenon will be further addressed in Sec. 4.4 below, where the a-PCF for the dilute swarm is discussed. The higher scatter of the experimental data for low void fractions suggest that for this regime the phenomenon is not so clear and further investigations are needed to clarify this point.

The slightly higher bubble velocity compared to the *SmMany* case is also the reason for the higher fluid velocity in the channel center for the dilute swarm *SmFew* (Fig. 4.3, left) compared to the denser swarms. The hindrance effect can also be the reason for the shape of the relative velocity profile of the bubbles in the *SmMany* case (Fig. 3.19) since reduced relative velocity correlates with increased bubble density.

Wall-normal profiles of bubble fluctuating velocities are presented in Fig. 4.3, right. For all components, they are smaller than for the denser swarm, confirming both the more regular bubble dynamics and the lower level of turbulence in the dilute swarm. Nevertheless, the fluctuations in the x-direction are higher than the ones in the z- and y-directions, the latter being limited by the walls and presenting the lowest values. A difference with respect to the denser swarm consists in the values of $\langle v'_p v'_p \rangle$ and $\langle w'_p w'_p \rangle$ at the channel center. For the *SmMany* simulation, the former is around twice as high as the latter, suggesting a higher bubble movement in the z-direction than in the y-direction for the bubbles in the denser

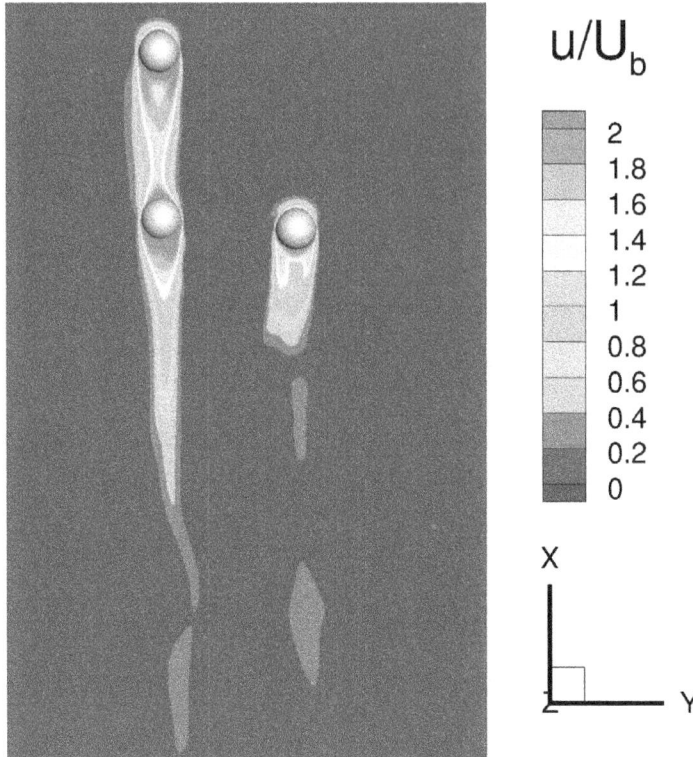

Figure 4.5: Respresentation of the wake-aspiaration mechanism for the *SmFew* case. The instantaneous fluid velocity profile is represented for an arbitrary instant in time.

swarm. For the case *SmFew*, both quantities have about the same value, and this confirms the isotropic fluctuations of bubbles and fluid in the horizontal directions in the center region, mentioned above.

As for the case *SmMany* above, the shape of the void fraction distribution is now addressed. The lift force moves the small bubbles toward the channel center, as discussed in Sec. 3.2.2 and 3.3.4, which alone does not explain the shape of $\langle \phi \rangle$, however. The number of collision events is even smaller than in the previous case so that collisions may be disregarded for bubble dynamics. The turbophoresis, instead, plays an important role for the bubble distribution. As for the case *SmMany*, the quantity T according to (3.20) and (3.21) was evaluated and is shown in Fig. 4.6. Both profiles of T present the same attitide and two regions can be observed:

- For $y/H \lesssim 0.13$, large negative values of T: Bubbles are moved toward the wall.

- For $y/H \gtrsim 0.13$, small positive values of T: Bubbles are moved toward the core region,

and, hence, the two regions separated by the zero crossing of T at $y_w/H \approx 0.13$ are very well defined. The region of slightly higher void fraction $\langle \phi \rangle$ is found in the interval $y_w/H \approx 0.12...0.25$. Its near-wall boundary $y_w/H \approx 0.12$ is very close to the position where T changes sign. As reported above, the lift force moves the bubbles toward the core region

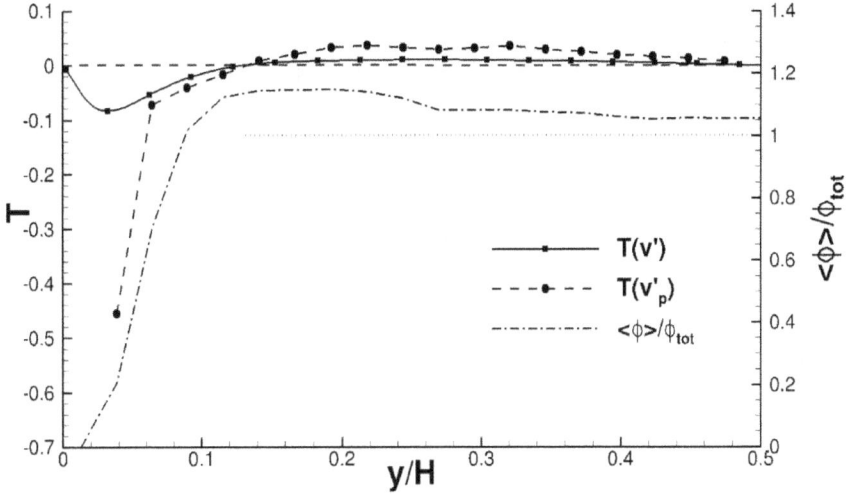

Figure 4.6: Quantification of the turbophoresis effect for simulation *SmFew*. Continuous line: Wall-normal profile of T according to (3.20) and (3.21) evaluated with central difference approximation. Dashed line: Void fraction distribution. Dotten line: $\langle \phi \rangle = \phi_{tot}$.

and, for $y/H < 0.13$, its effect dominates over the turbophoresis, yielding the distribution observed[1]. Compared with the denser swarm *SmMany*, the influence of the turbophoresis on the void fraction distribution does not seem so substantial and this impression is backed by the smaller magnitude of T for the dilute swarm. When evaluated with the fluctuations of the fluid velocity, the smallest value of T for the *SmFew* case is about one third of the smallest value of T for the *SmMany* case only. On the other hand, the shear rate of the fluid velocity near the wall is close to the one of *SmMany* (Fig. 3.17), so that the lift force on the individual bubble is very much the same for the two cases. This allows the lift force to dominate for *SmFew*, while it is the turbophoresis for *SmMany*. In conclusion, even if the bubble distribution results from the complicated balance of all terms in (3.19), the turbophoresis effect can explain the observed behavior of the void fraction distribution.

4.4   Pair correlation functions

The r-PCF according to (3.22) for the dilute swarm is presented in Fig. 4.7. The most probable pair distance is around $6r_p$, which is larger than in the denser swarm. The lower the global void fraction, the higher the MPPD. In a dilute swarm in pseudo-turbulence, Martinez Mercado et al. (2010) found an MPPD of 4 equivalent radii for void fraction between 0.28% and 0.74%. The difference can be explained by the role of turbulence in the present investigation, which increases the bubble agitation, yielding larger distances between the bubbles. For the investigated regime, these authors did not find smaller MPPD with increasing void fraction. As in the denser swarm *SmMany*, three regions can be found

---

[1]Note that if only the shear-induced lift force would act on the bubbles, a parabolic-like profile of $\langle \phi \rangle$ can be expected, with very few bubbles in the wall-region and many more in the core region.

when comparing the r-PCF of the *SmFew* case with the one of RDB. In the first region, $2 \lesssim r_p/H \lesssim 5$, pairs are less probable than in the random case. For $5 \lesssim r_p/H \lesssim 15$ pairs are more probable. For $r_p/H \gtrsim 15$, pairs are again less probable than in the random case.

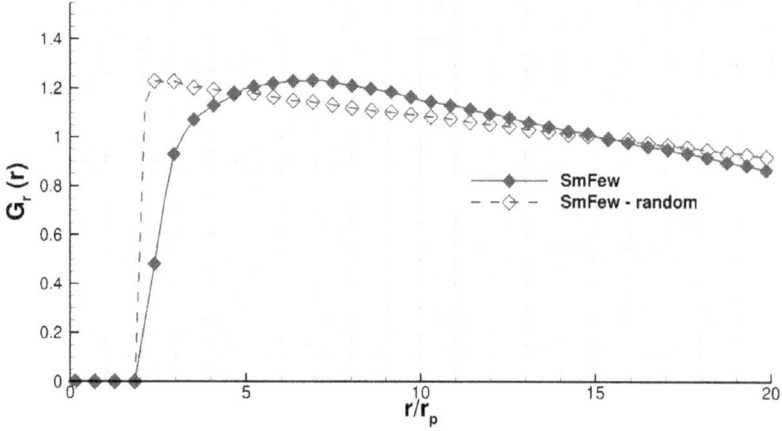

Figure 4.7: Radial pair correlation function, r-PCF, for *SmFew* and RDB.

The a-PCF for the dilute swarm is shown in Fig. 4.8 in form of the canonical one-dimensional curves and by means of a two-dimensional representation. For small distances, bubbles align mostly vertically. This can be explained by the the wake effect experienced by a trailing bubble and the so-called aspiration mechanism. At small distances, the local maximum for $\phi_g = 90°$ is due to the lateral pressure mechanism (see Sec. 3.3.5) which takes place when two bubbles happen to rise near each other. For increasing distances, two features are observed. The first concerns the relative reduction of the horizontal alignment for increasing pair distances. This is due to the pressure-related mechanism that becomes weaker for increasing distances and is in accordance with the experimental results of Martinez Mercado et al. (2010) for pseudo-turbulence, see Fig. 7 therein. The second observation for increasing pair distances is that the values of the maxima for the vertical alignment first increase and then decrease. Additionally, the position of the minima is shifted for the different shells. This observation can be related to the very few statistical samples of bubble pairs for very short distances: For each investigated field, on average only 3.6 pairs are detected for $d_{i,j} \leq 3r_p$, yielding an overall amount of $\mathcal{O}(100000)$ pairs considering all bubble fields analyzed spanning the whole simulation time. For the denser swarm *SmMany*, the number of bubble pairs found for the smaller distance investigated $d_{i,j} \leq 3r_p$ is around 100 times higher, with 296 pairs detected, on average, for each bubble field.

The comparison with the denser swarm and the results regarding the hindrance effect provide the following picture. In a dilute swarm, bubbles tend to follow the vertical path of the leading ones. This is also due to the limited fluctuations of the particle velocity in the horizontal plane (Fig. 4.3, right). This feature is confirmed by the preferential vertical alignment of bubbles for small distances and by the slightly higher mean rise velocity, since a trailing bubble in the wake of a leading one rises faster, eventually reaching the leading one, according to the "drafting, kissing and tumbling" mechanism (see also Fig. 4.5). The lateral pressure mechanism, as discussed in Sec. 3.3.5, is a short-range effect and, hence, does not play an important role in a dilute swarm, since the distances between the bubbles

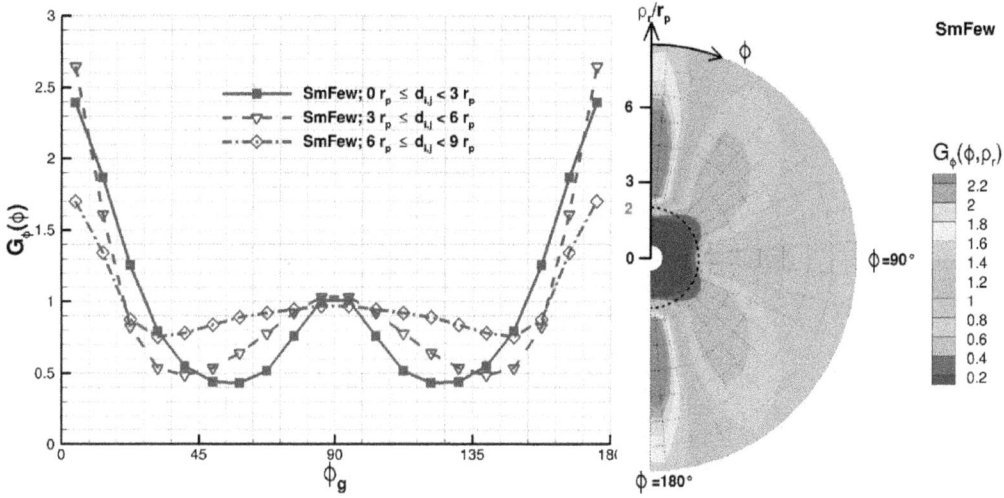

Figure 4.8: Angular PCF for *SmFew*. Left: Canonical representation. Different curves for $d_{i,j}/r_p \in [0;3[$, $d_{i,j}/r_p \in [3;6[$ and $d_{i,j}/r_p \in [6;9[$. For this picture $r_2 - r_1 = 3\,r_p$ in (3.23). Right: Two-dimensional representation, with $\rho_r = (r_2 - r_1)/2$ and $r_2 - r_1 = r_p$.

are larger. Within the denser swarm, turbulence is enhanced, bubble fluctuations are higher and the leading bubble/trailing bubble constellation is not the dominant one. For such a swarm, the reason for horizontal alignment is the lateral pressure field since bubble-bubble distances are much smaller.

In conclusion, the comparison between a dilute and a denser swarm can be summarized as follows. Regarding the bubble-induced flow structures, traces of the usual single-phase flow structures are still found in the channel, but are now influenced by larger structures related to the bubble presence. Such structures are smaller than the corresponding ones observed in the denser swarm. Turbulence is enhanced with respect to the unladen flow but not so much as for the denser swarm since the bubble-induced turbulence is related to the total void fraction. The hindrance effect was marginally observed in the dilute swarm and found in good agreement with experimental results for the denser swarm and the turbophoresis is also dominant over the lift force for the mean void fraction distribution. Bubbles in the dilute swarm align mainly vertically for small distances due to the aspiration mechanism of the wake of a leading bubble.

5 Influence of bubble size

When the dynamics of a single bubble in quiescent fluid was discussed in Sec. 3.2.1, it was mentioned that, even in this simple situation, many flow features such as the bubble paths and the structure of the wake strongly depends on the bubble Reynolds number. Many studies have been devoted to this topic and a detailed review is provided in (Ern et al., 2012). One may now aske the question, whether such differences are also found for bubbles in a turbulent channel flow and how the expected different behavior influences the surrounding fluid. Strictly connected to the influence of the bubble size is also the influence of the polydispersity of the bubble swarm, which represents a situation more similar to the bubbly flows occurring in real configurations. Some investigations have already been performed to address these topics, for example by Lu and Tryggvason (2013), who investigated the effect of one large deformable bubble on the dynamics of a swarm of small bubbles in turbulent channel flow, compared to a swarm consisting only of small bubbles. These authors observed that the large bubble, rising mainly in the channel center, increased the turbulent kinetic energy in the core region due to the production of vorticity at the phase boundary. Furthermore, it was observed that the presence of the large bubble modifies the clusters of small bubbles at the wall reducing, for example, the tendency of small bubbles to align horizontally. Unfortunately, such comparison was not performed in the core region, where the large bubble rises. Göz and Sommerfeld (2004) investigated the rise of bidisperse swarms in otherwise quiescent fluid in a triple periodic domain varying the total void fraction and the ratio of the volumes of the two size classes. The liquid turbulence induced by the bubbles as well as the tendency of bubble pairing were analyzed and it was observed that the presence of second class of bubbles can modify the behavior of the bubbles when compared to monodisperse swarms. In the framework of Euler-Euler simulations, Politano et al. (2003) employed a polydisperse approach combined with the standard $k - \epsilon$ model to investigate the influence of bubble size on the flow feature of upward bubbly flows. Similarly, Krepper et al. (2005, 2008) developed a so called "inhomogeneous multiple size group" model to account for many classes of bubbles in the framework of Eulerian modeling.

This brief overview highlights the need to perform trustworthy simulations of bubbly swarms to address the role of the bubble size and of the polydispersity of the swarm, since it is expected that such parameters may have a large influence on both the fluid and the bubble dynamics. The two simulations presented here will therefore contribute to throw light on the complex phenomena involved in this type of flows. To this end, a simulation labeled *LaMany* is performed under the same conditions employed for the *SmMany* case where the swarm is constituted by larger bubbles. Additionally, a second simulation labeled *BiDisp* is considered where a bidisperse swarm is investigated, where half of the total void fraction is constituted by small bubbles and the other half by larger bubbles.

This chapter is organized as follows. First, the two reduced problems addressed in Sec. 3.2 are repeated for the analysis of large bubbles: The dynamics of a single bubble rising in quiescent fluid and the response of a fixed bubble in cross-flow with a constant shear rate These simulations, presented in Sec. 5.1, serve both as validation for the numerical code and to obtain reference data that will be employed when the dynamics of large bubbles in the swarms is investigated. Afterward, the simulation *LaMany* is presented in Sec. 5.2 and compared to the simulation *SmMany* to investigate the role of the bubble size on the channel flow. The ratio between the diameters of the large and the small bubbles is 1.46 (see Table 3.2) and, though relative small, this value is large enough to observe significant differences, as will be reported below. Finally, in Sec. 5.3, this jointed analysis is completed introducing the results for the *BiDisp* case and addressing the role of the bidispersity.

5.1   Validation for large bubble cases

5.1.1   Large bubble in quiescent fluid

For the rise of a large bubble the computational domain is a rectangular column whose extensions are $232.7\,d_p \times 11.3\,d_p \times 11.3\,d_p$ in the x-, y- and z-direction, respectively and periodic conditions are applied on the six boundaries. This simulation is labeled *SingleLa* in the following. The bubble is discretized by around 17 Eulerian points across the diameter, i.e. $d_p/\Delta = 17.1$, as for the large bubbles in the channel flow, see Tab. 3.2. The density ratio ρ_p/ρ is 0.001. Compared with the simulation presented in Sec. 3.2.1, the domain is somewhat longer and narrower. Figure 5.1 portrays the time history of the velocity components: For $t/t_{ref} > 10$, the streamwise component u_p is fairly constant but exhibits reduced values when the v_p-component exhibits its maximal values for $t/t_{ref} \approx 50$ and $t/t_{ref} \approx 100$.

The u_p-component is averaged over more than $100 t_{ref}$ and, based on its average value $u_{p,\infty}$, a bubble Reynolds number according to (3.2) is obtained and its value is $Re_{p,\infty} = 497$ for the present simulation. Based on this value, the bubble dynamics can be compared to the work of Horowitz and Williamson (2010) which investigated experimentally the rise and the fall of a sphere in quiescent fluid. Due to the spherical shape and the no-slip condition at the phase boundary, the analysis in (Horowitz and Williamson, 2010) is best suited for comparison, as mentioned in Sec. 3.2.1. For a sphere at $Re_{p,\infty} = 450$ and $\rho_p/\rho = 0.08$ these authors observed an oscillatory path. Figure 5.2, left, portrays the bubble trajectory projected onto the horizontal plane yz and features of an oscillatory path can be observed. The bubble trajectory is therefore in good agreement with the observation in (Horowitz and Williamson, 2010). Nevertheless, the somewhat irregular oscillations of the velocity components observed in Fig. 5.1 can also be observed in the projection of the path and hint at a trajectory with features from both an oscillatory and a chaotic path. The wake structure is portrayed in Fig. 5.2, right, with two different perspective views. Bent hairpin-like vortex detached from the bubble surface are observed and are characteristic for oscillatory paths, as described in Horowitz and Williamson (2010) and Ern et al. (2012).

Based on $u_{p,\infty}$, a mean drag coefficient can be evaluated and for the present simulation the value is $C_D = 0.81$. This value is in very good agreement with the value evaluated in Horowitz and Williamson (2010) for "vibrating" sphere, see Fig. 3.4 in Sec. 3.2.1.

It is now recalled that for the small bubble at $Re_{p,\infty} = 265$ investigated in Sec. 3.2.1, a slight oblique path was observed with limited oscillations and different, somewhat more irregular wake structures are observed (cf. Fig. 5.2 and Fig. 3.3). These observations and the analysis of the trajectories allow stating that the dynamics of small and large bubbles addressed below in the channel flow is expected to be quite different. Indeed, the dynamics of large bubble is much more complex than the one of small bubbles, inducing a higher agitation of the surrounding fluid. This issue will be addressed and quantified below when the two monodisperse swarms are compared.

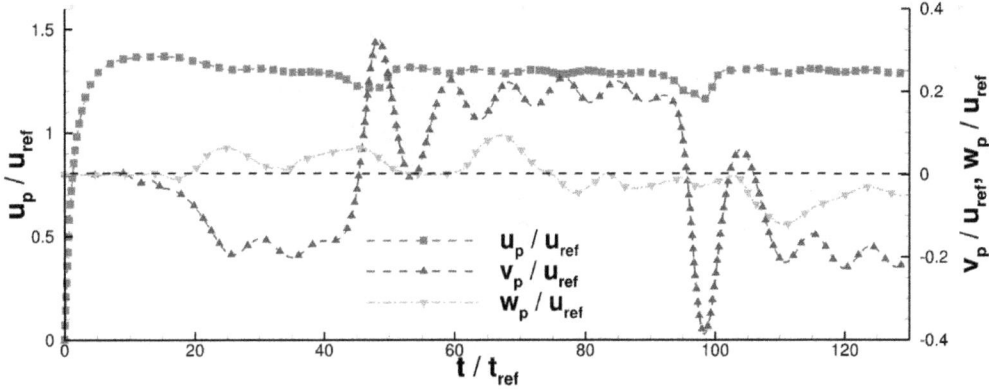

Figure 5.1: Time history of bubble velocity components for the case *SingleLa*. The straight dashed line represents the zero value of the right ordinate.

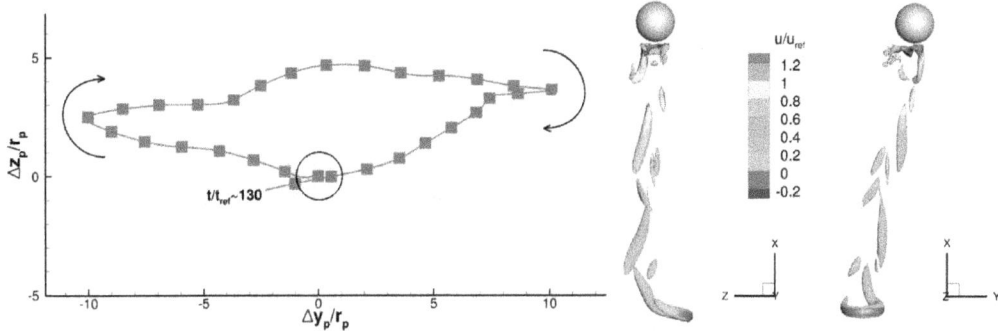

Figure 5.2: Left: Top-view (zy) of bubble path for case *SingleLa*. The size of the bubble is marked by a circle around the start of the trajectory at $t/t_{ref} = 0$. The end of the recorded trajectory is reached at a vertical position of about $220d_p$ after a time of $130t_{ref}$. Right: Flow visualizations for case *SingleLa* by two-perpendicular views of the same situation: Isosurface of λ_2, colored with the instantaneous fluid velocity in the x-direction. The λ_2 isosurface was chopped at the rear end of the bubble not to cover the bubble itself.

5.1.2   Large bubble in cross-flow

As for the small bubble at lower Re_p, the response of a fixed large bubble in a cross-flow with constant shear rate is investigated. To this end, a simulation labeled *FixedLa* is conducted where a fixed large bubble is considered. For this case the computational domain is defined as $[0; 17d_p] \times [0; 8.5d_p] \times [0; 8.5d_p]$ in the stream-wise, shear- and span-wise direction, i.e. x-, y-, and z-direction, respectively. At the inlet a linear velocity profile is imposed, with dimensional shear parameter $S = (\partial u / \partial y) r_p / U_c = 0.1$, where U_c is the velocity at midspan. The sphere is positioned at $(x/d_p, y/d_p, z/d_p) = (4.25, 4.25, 4.25)$ and the Reynolds number $Re_{p,c}$ is set to 420, which was the first guess of the Reynolds number of large bubbles in the channel investigated below. The ratio between particle diameter and mesh step size is around 17 as in the channel flow configuration, yielding a total number of grid points of around 6.7 Million.

Figure 5.3 shows a lateral view and a top view of the sphere and of the wake structures represented by isosurfaces of the vorticity in the streamwise direction $\omega_x \pm 4U_c/d_p$ after a oscillatory steady stated s reached. Additionally, the history of the force coefficient C_y in the y-direction according to (3.4) is portrayed, where the simulation is scaled with $t_{ref} = d_p/U_c$. Two features can be observed. The first is that C_y is always negative, indicating that the lift force is directed towards the side of the sphere with the lower relative velocity throughout, as for the small bubble. Based on the stationary oscillatory state, the mean values $\langle C_y \rangle_t$ is -0.071, varying between -0.059 and -0.083, and this value is in accordance with the results in Kurose and Komori (1999) portrayed in Fig. 3.6. The second feature is that both the flow visualizations and the history of C_y hint at a periodic behavior of the wake, with periodic vortex shedding from the upper side of the sphere. This is the side of the bubble where the relative velocity between fluid and particle is higher. Oscillation of the lift force for $Re_{p,c} > 300$ was observed also in (Kurose and Komori, 1999).

The lift force, furthermore, is directed towards the side of the sphere where the shedding does not take place. This observation agrees with the results of Bagchi et al. (2001) for a fixed sphere at somewhat lower $Re_{p,c}$ in uniform cross flow. For $Re_{p,c} = 350$ and without shear, these authors found that symmetry breaking occurs and one-sided shedding takes place. This leads to a non-zero time averaged lift force which is directed towards the side of the sphere with no shedding. The present analysis, hence, shows that one-sided vortex shedding also occurs in linear shear flow and that an average lift force towards the other side is generated. Furthermore, according to the present study, the relation between one-sided vortex shedding and the direction of the lift force was observed also for bubbles in shear flow.

As for the small bubbles, the shear-induced lift force is supposed to move the large bubbles toward the channel center, i.e. toward the side of the sphere where the relative velocity is lower. This phenomenon will be further investigated below where the distribution of the large bubbles in the channel is addressed.

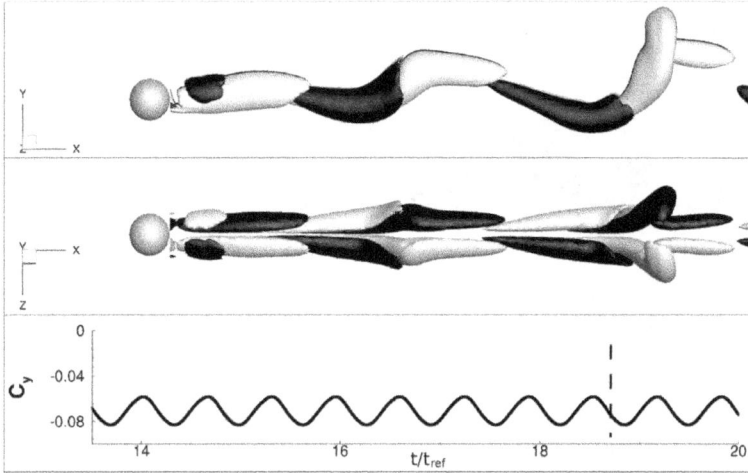

Figure 5.3: Wake structure behind a fixed sphere in shear flow for simulation *FixedLa*. The wake is represented by isosurfaces of the vorticity in the streamwise direction $\omega_x = \pm 4U_c/d_p$, colored with the instantaneous fluid velocity in the z-direction. The velocity gradient is oriented in the y-direction. The vorticity isosurfaces were chopped at the rear end of the bubble not to cover the bubble itself. Top: side view. Middle: top view. Bottom: Time history of the lift coefficient C_y according to (3.4). The vertical dashed line represent the instant of the flow visualizations above.

5.2 Results for the monodisperse swarm of larger bubbles

In this section the analysis of the flow laden with a swarm of large bubbles, labeled *LaMany*, is presented. Comparisons are drawn with the corresponding quantities of the unladen flow, case *Unladen*, and of the swarm laden with small bubbles, case *SmMany*, both analyzed in Sec. 3.3. This jointed analysis allows, hence, addressing the influence of the bubble size on the fluid field and on the bubble dynamics.

5.2.1 Fluid phase: Flow structures and correlation functions

A first impression of the instantaneous flow is provided in Fig 5.4, where three-dimensional velocity fluctuations according to (3.5) are shown. For the threshold value employed, one large elongated structure can be observed, spanning the whole channel in the streamwise direction. With respect to the instantaneous structures in the *SmMany* case, in Fig. 3.7, right, the size appears to be longer and wider in the present swarm with larger bubbles. The turbulence level associated with the swarm of larger bubbles is also higher and can be appreciated by the roughness of the portrayed structures. Near-wall structures are not found with the employed fluctuation value, since their turbulence lever is much lower than the one of the large structures. The two-point correlation function in the streamwise direction according to (3.6) shows an identical behavior in the near-wall region and in the channel center, Fig. 5.5 and Fig. 5.6, respectively. For clarity the curves of the *Unladen* case are not reported here and the reader may refer to Sec. 3.3.1. For small scales, the reduction

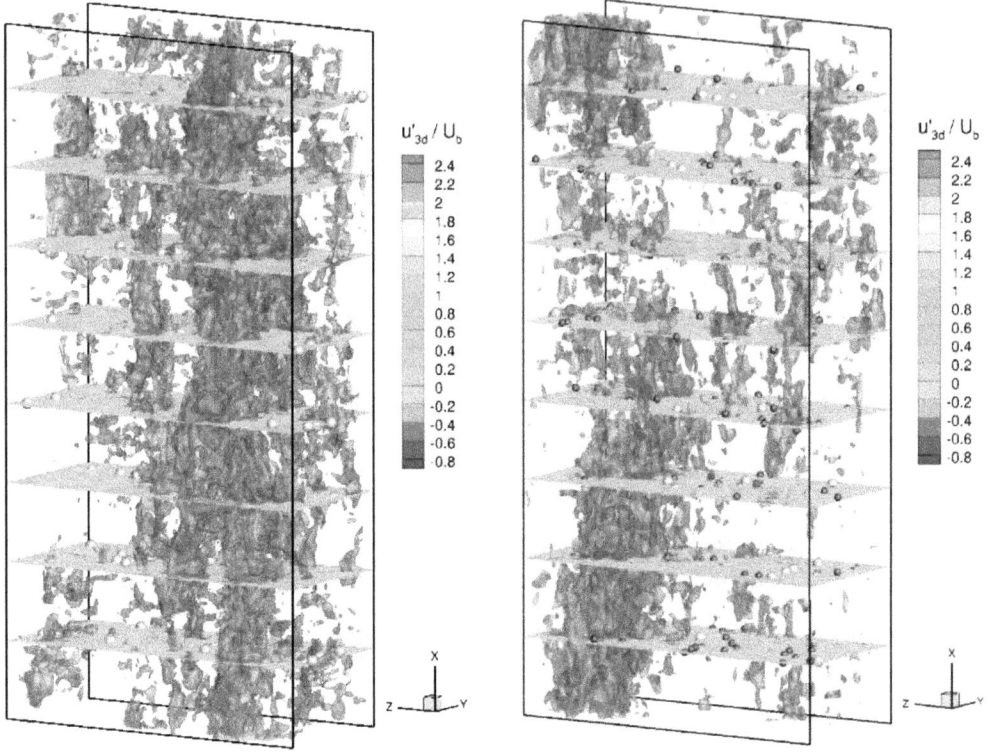

Figure 5.4: Instantaneous three-dimensional velocity fluctuations for simulation *LaMany*, left, and *BiDisp*, right. Iso-surfaces of $u'_{3d}/U_b = -0.5$ according to (3.5); The vertical distance between the planes used for color plots is $0.5H$. Only bubbles cutting such planes are represented.

of correlation with respect to the *Unladen* case is confirmed due to turbulence production associated with bubbles at the small scales. For large scales, the correlation is always positive and higher than both in the *Unladen* case and in the *SmMany* case. This confirms, as for the the *SmMany* case, that the long, elongated structure depicted in Fig. 5.4 is a statistical feature. The different large-scale behavior of the flow for the *SmMany* case and for the *LaMany* one is hence confirmed by both flow visualizations and by a statistical analysis. In the flow laden with small bubbles, $R_{uu} < 0$ for $\Delta x/H \approx L_x/2$ suggested the presence of regions of opposite fluctuation sign at such distance, as if a region of positive fluctuations is followed by a region of negative fluctuations, even if the negative value of R_{uu} is quite small. For the *LaMany* case, instead, $R_{uu} > 0$ for $\Delta x/H \approx L_x/2$ implies that there is one single structure, i.e. one region where the fluctuations presents the same sign, as portrayed in Fig. 5.4. This indicates, for the investigated configuration, that a dense swarm of large bubble induces flow structures which are longer than $4.43H$, i.e. $117r_p$. As discussed in Sec. 3.3.1, such long structures are somehow typical of channel flow laden with finite-size objects (bubbles or particles), both for co-current flows (like the ones investigated in the present study) and counter-current flows (Uhlmann, 2008; Garcia-Villalba et al., 2012).
The instantaneous two-dimensional velocity fluctuations according to (3.9) for an arbitrary

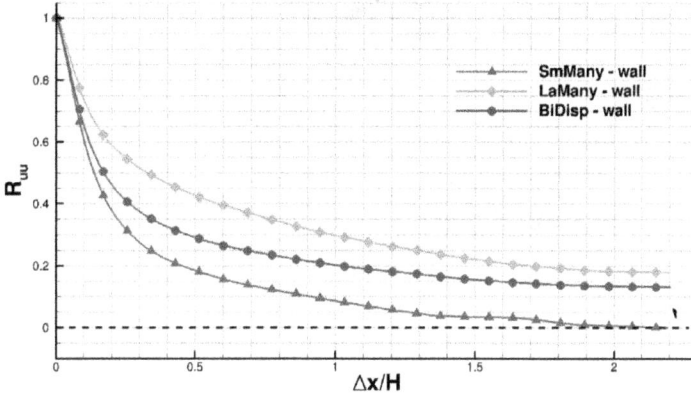

Figure 5.5: Two-point correlation function R_{uu} according to (3.6) in the near-wall region $y \approx 0.02H$.

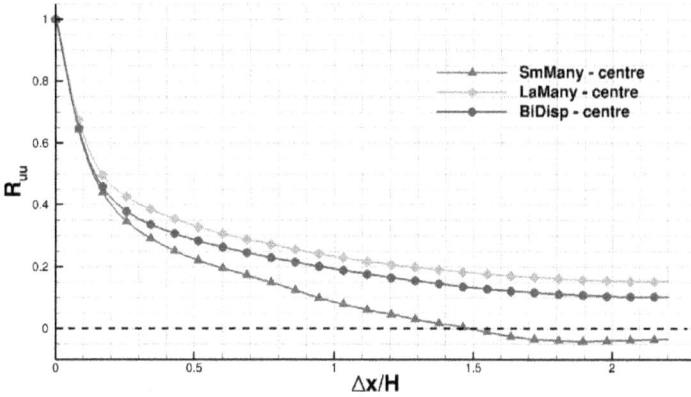

Figure 5.6: Two-point correlation function R_{uu} according to (3.6) in the centre region.

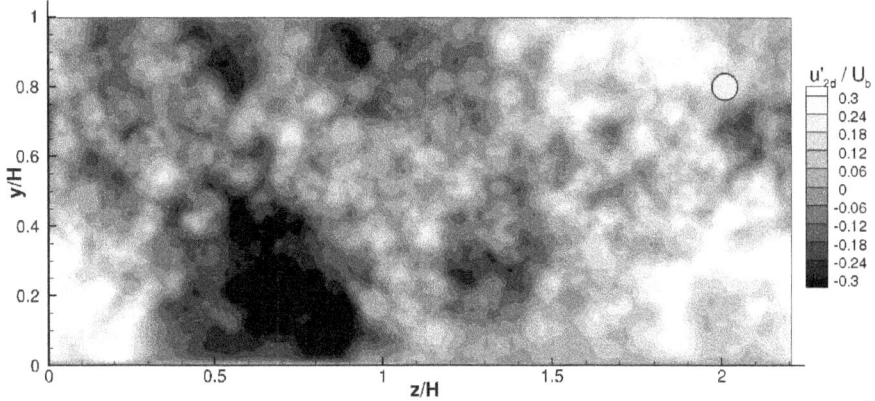

Figure 5.7: Instantaneous two-dimensional velocity fluctuations u'_{2d} according to (3.7) for simulation *LaMany*. The bubble size is represented by a circle at $y/H = 0.8$, $z/H = 2$.

instant in time are shown in Fig.5.7. Two regions of almost equal size can be observed, one related to positive fluctuations and the other to negative fluctuations. The region of negative fluctuations consists of one medium-sized structure (centered at $y/H \approx 0.2$ and $z/H \approx 0.7$) and three small-size structures, while the region with positive fluctuations consists of two structures of about the same extension.

The instantaneous streamwise-averaged bubble concentration $N_{b,2d}$, according to (3.8) is shown in Fig. 5.8 for the flow field portrayed in Fig. 5.4 and 5.7. The bubbles obviously concentrate in the center of the channel with very low void fraction near the walls. This will be further discussed in Sec. 5.2.3 below. Hence, it can be observed that slightly higher void fraction seems to coincide with somewhat larger u'_{2d}-regions, e.g. around ($y/H \approx 0.4$, $z/H \approx 2$) and vice versa around ($y/H \approx 0.1$, $z/H \approx 0.6$). Nevertheless, the relation between positive and negative fluctuation regions of spanwise fluid structure and bubble concentration in the plane is not straightforward. A statistical quantification of the flow behavior in the spanwise direction is provided by the two-point correlation function in the spanwise direction, defined analogously to (3.6). In the near-wall region shown in Fig. 5.9 no sign of small-scales near-wall structures can be discerned. The correlation is negative for large distances so that positive and negative fluctuations occur alternating, on average. In the channel center, Fig. 5.10, this feature is found as well and is substantially more pronounced. Remarkably, R_{uu} is identical for all three cases beyond $\Delta z/H = 0.35$, which corresponds to around $4.4 d_p$ for the large bubbles and $6.7 d_p$ for the small ones (see Tab. 3.2).

5.2.2   One-point statistics of the carrier phase

For the swarm of large bubbles, the time-averaged wall-shear stress, collected in Table 3.3, is slightly higher than in the unladen flow but lower with respect to the swarm of small bubbles, case *SmMany*. The fluctuations around the mean value (Tab. 3.3) are higher with respect to the other simulations and this can be related to the presence of large flow structures which approach the walls and induce a higher shear stress.

The mean fluid velocity (Fig. 5.11) is strongly modified by the swarm of large bubbles. The

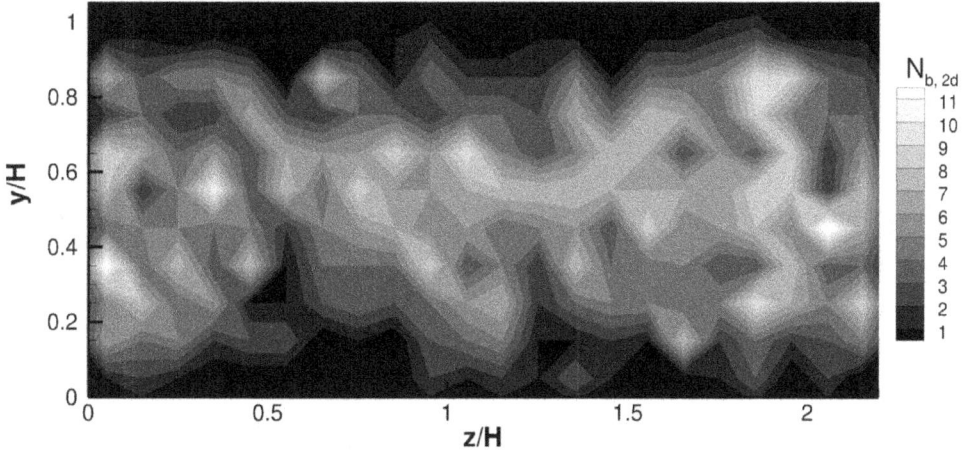

Figure 5.8: Two-dimensional distribution of bubbles in the yz-plane according to (3.8) for the flow field depicted in Fig. 5.7.

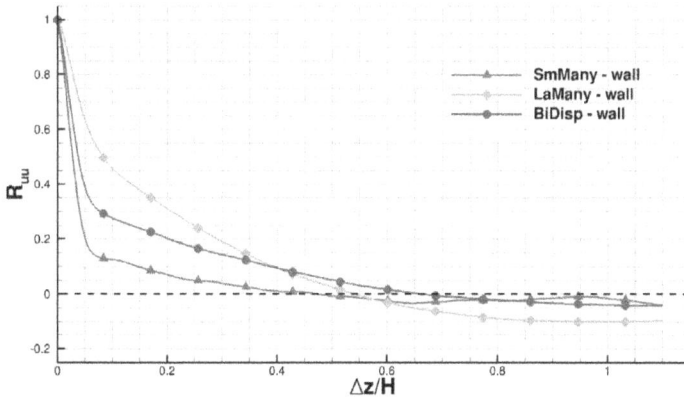

Figure 5.9: Two-point correlation function R_{uu} in the spanwise direction, according to (3.6), in the near-wall region $y \approx 0.02H$.

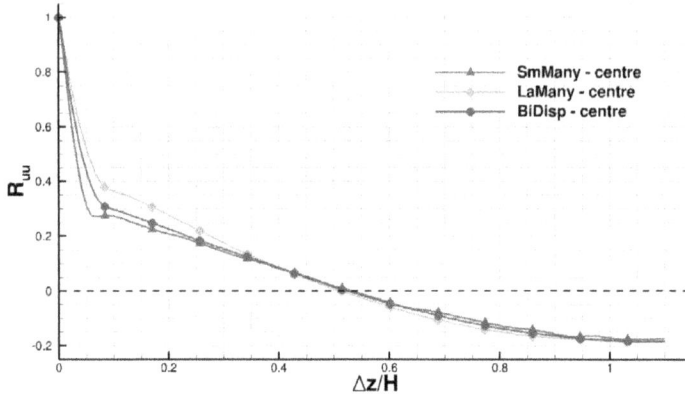

Figure 5.10: Two-point correlation function R_{uu} in the spanwise direction, according to (3.6), in the center region.

shape deviates from the turbulent velocity profile in channel flows, with higher values in the center region, due to the higher bubble concentration in the channel center (Fig. 5.8). The velocity profile is very slightly asymmetric, with slightly higher values for $y/H > 0.5$. When portrayed in wall units, Fig. 5.11 right, the mean velocity profile falls onto the one of *SmMany* up to $y^+ \approx 20$. Beyond, the increase with y^+ is much stronger and no logarithmic region can be observed. This is due to the strong vertical forcing induced by the bubbles, which are not uniformly distributed as shown below.

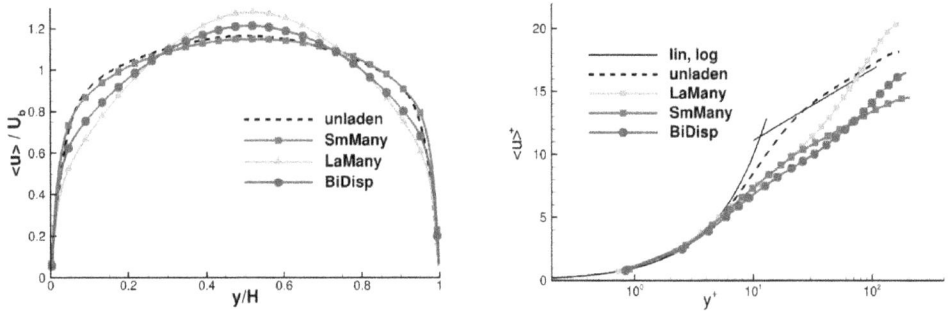

Figure 5.11: Averaged streamwise velocity for *SmMany*, *LaMany*, *BiDisp* and *Unladen*. Left: bulk units; right: wall units.

The non-vanishing components of the Reynolds stress tensor are shown in Fig. 5.12. They still exhibit a certain similarity in shape with the ones of the unladen flow and of the simulation with small bubbles show substantial quantitative differences. The streamwise normal stress, $\langle u'u' \rangle$, reaches up to a maximum of $0.21U_b^2$, while the maximum is $0.03U_b^2$ for the unladen case and $0.11U_b^2$ for *SmMany*. Furthermore, the position of the maximum is shifted toward the center of the channel and now is observed at $y/H = 0.23$, $y^+ \approx 80$ while these values are $y/H = 0.04$, $y^+ \approx 13$ for the unladen flow and $y/H = 0.05$, y^+6 for the case *SmMany*. Compared to the case *SmMany*, where $\langle u'u' \rangle$ exhibits a broad plateau in the center region, $\langle u'u' \rangle$ is much smoother and the area of larger values much broader.

The spanwise stress, $\langle w'w' \rangle$, shows a similar attitude. The maximum is attained at the same distance from the wall and the shape is broader than for *Unladen* and *SmMany*. The wall-normal fluctuations, $\langle v'v' \rangle$, increase smoothly from the wall and have no local maxima but a plateau over the center region with maximum 0.044 compared to 0.001 for *Unladen* and 0.009 for *SmMany*. The turbulent shear stress, $\langle u'v' \rangle$, inherits the smooth shape of $\langle u'u' \rangle$ and $\langle v'v' \rangle$ and does no more exhibit a constant slope in the center region as the the unladen channel and the case *SmMany*.

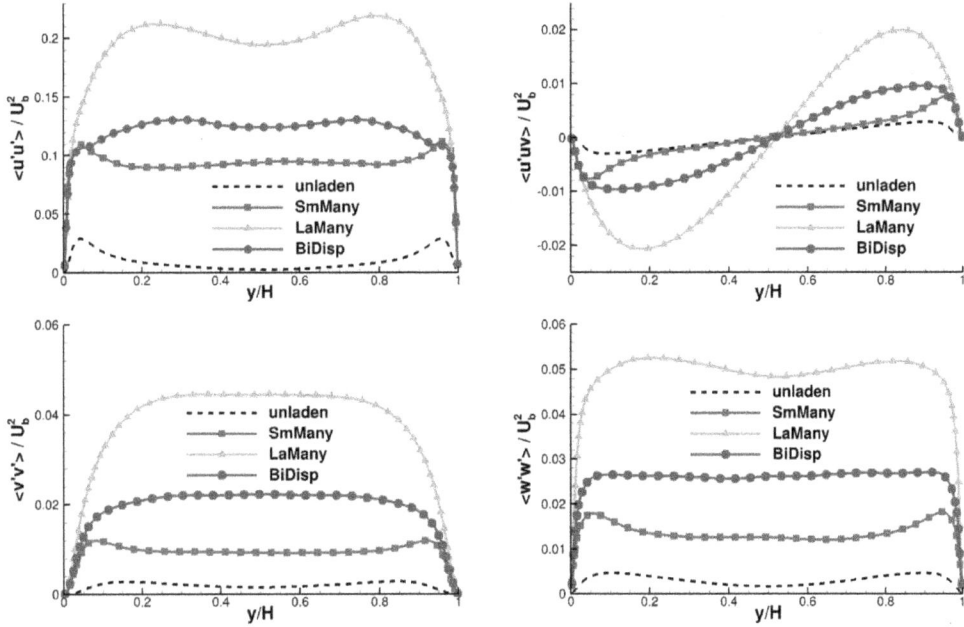

Figure 5.12: Reynolds stress components for *SmMany*, *LaMany*, *BiDisp* and *Unladen*. Top, left: streamwise fluctuations $\langle u'u' \rangle$; top, right: turbulent shear stress $\langle u'v' \rangle$; bottom, left: wall-normal fluctuations $\langle v'v' \rangle$; bottom, right: spanwise fluctuations $\langle w'w' \rangle$.

The enhancement of the turbulence level is in line with the quantification proposed by Hosokawa and Tomiyama (2004) based on the ratio R between the eddy viscosity induced by the disperse phase and the shear-induced eddy viscosity according to (3.12). For the present simulation, equation 3.12 yields R=5.34 which is almost twice as the value of R for the *SmMany* case.

5.2.3 One-point statistics of the disperse phase

As for the *SmMany* case, for the wall-normal statistics of the bubbles the width of the wall-normal bin was chosen equal to the bubble radius, namely $I_w = r_p$. The storage frequency of bubble data was modified accordingly accounting for the mean rise velocity of large bubbles, so that still the bubbles travel on average one bubble diameter between the instances in time when bubble data are stored for later averaging.

The mean bubble distribution $\langle \phi \rangle$, as depicted in Fig. 5.13, left, shows that large bubbles rise, on average, more frequently in the channel center. This is related to the direction of the shear-induced lift force, which moves large bubbles toward the channel center (as will be discussed below). The tendency of large bubbles to rise in the center region explains the shape of the fluid velocity profile shown in Fig. 5.11. Since they rise more frequently in the channel center, the fluid, by this uneven distribution, is entrained more in the center region so that the vertical velocity is larger in this region. The very slight asymmetry of the profile of $\langle \phi \rangle / \phi_{tot}$ is also the reason why the profile of $\langle u \rangle / U_b$ is slightly asymmetric. When compared to the small bubbles of the *SmMany* case (Fig. 3.19), the rise velocity and, hence, the relative velocity, portrayed in Fig. 5.13, left, are larger for larger bubbles than for small ones, as expected due to larger buoyancy force. The larger relative velocity in the wall regions is due to the local hindrance effect and can be related to a local reduction of the hindrance effect, since the void fraction is smaller so that the distance between the bubbles is larger (as in the *SmFew* swarm addressed in Sec. 4.3). The shape of the relative velocity is in contrast with the one of the *SmMany* case, since for the latter the void fraction exhibits a slight increase toward the wall resulting in a slight decrease of u_r due to the increased hindrance.

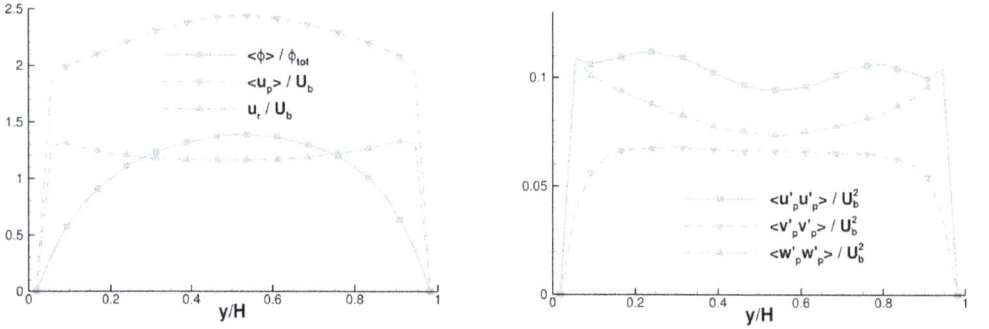

Figure 5.13: Left: Void fraction distribution, bubble velocity and relative velocity for case *LaMany*. Right: Bubble velocity fluctuations for case *LaMany*.

Fluctuations of the bubble velocity components are shown in Fig. 5.13, right. Among all components, the streamwise one, $\langle u'_p u'_p \rangle$, presents the highest values. The spanwise component exhibits comparable values near the wall, but lower values in the core region. As in then *SmMany* case, both the streamwise and the spanwise fluctuation components present a maximum at the wall, in contrast to the wall-normal fluctuations which increases smoothly away from the wall and have an almost uniform value for for $0.2 < y/H < 0.8$. For all components a slight asymmetry is present, with the fluctuations slightly higher for $y/H < 0.5$. The position of the maxima of the streamwise component $\langle u'_p u'_p \rangle$ at around $y_w/H \approx 0.22$ corresponds to the position of the maxima of the streamwise velocity fluctuations $\langle u'u' \rangle$, as shown in Fig. 5.12. This indicates a strong correlation between fluid perturbations and bubble dynamics.

When compared with the small bubbles, the velocity fluctuations of the large bubbles are higher for all components and this observation holds also when the velocity fluctuations are scaled with the rise velocity $\langle u_p(y) \rangle$ and with the relative velocity $\langle u_r(y) \rangle$, as shown in Fig.

5.14 and 5.15, respectively. It can be appreciated that the fluctuation level is at least twice as large for the large bubbles when compared to the small ones, even when the different rise velocities are employed for the scaling. This observation backs the impression that the different fluctuation level is due to the different bubble paths, and due to the tendency of large bubbles to follow a zigzag path due to the higher Reynolds number. Different paths and different wake structures were also observed for bubble rising in quiescent fluid at different bubble Reynolds number (see Sec. 3.2.1 and Sec. 5.1.1).

The different agitation induced in the fluid by small and large bubbles can also ob observed by the turbulent shear stress $\langle u'v' \rangle$ presented in Fig. 5.12. If only the region $0.3 \leq y/H \leq 0.7$ is considered, the linear slope of $\langle u'v' \rangle$ is increased by a factor 7.5 with respect to the *SmMany* case. This enhancement is due to the increased turbulence level induced by the large bubbles but also related to the bubble paths. An oscillatory path yields displacements of the surrounding fluid in the vertical and in the horizontal direction that are strongly correlated. Hence, a stronger correlation between the u- and the v-component of the fluid velocity is observed and the linear slope of $\langle u'_p v'_p \rangle$ is increased by factor 6.5 between case *SmMany* and case *LaMany*, as reported in Fig. 5.16.

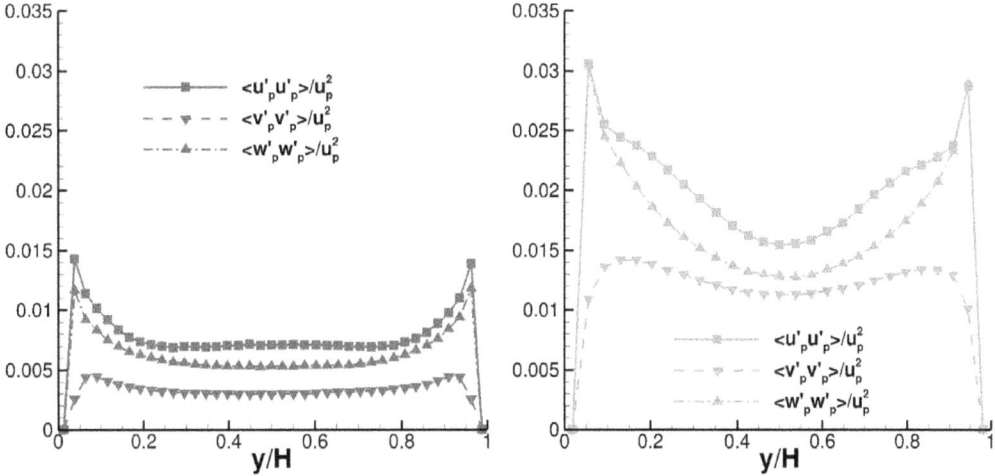

Figure 5.14: Fluctuations of bubbles velocity components when normalized with the bubbles rise velocity for the *SmMany* case (left) and for the *LaMany* case (right).

Regarding the averaged bubble distribution, it has been mentioned before that larger bubbles mainly rise in the core region and this is due to the shear-induced lift force (as discussed in Sec. 5.1.2). The turbophoresis effect, which accounts for the tendency of bubbles to migrate preferentially from regions of higher turbulence level toward regions of lower turbulence level, was also investigated and quantified by T according to (3.20) as reported in Sec. 3.3.2. For the present swarm consisting of large bubbles, T is always negative and moves the bubbles toward the walls. Nevertheless the lift force induced by the mean shear flow is the dominant effect so that the turbophoresis plays a marginal role for large bubbles.

Lift force on small and on large bubbles. As observed in (Kurose and Komori, 1999) and confirmed by preliminary analysis the lift force on a fixed bubble is directed toward

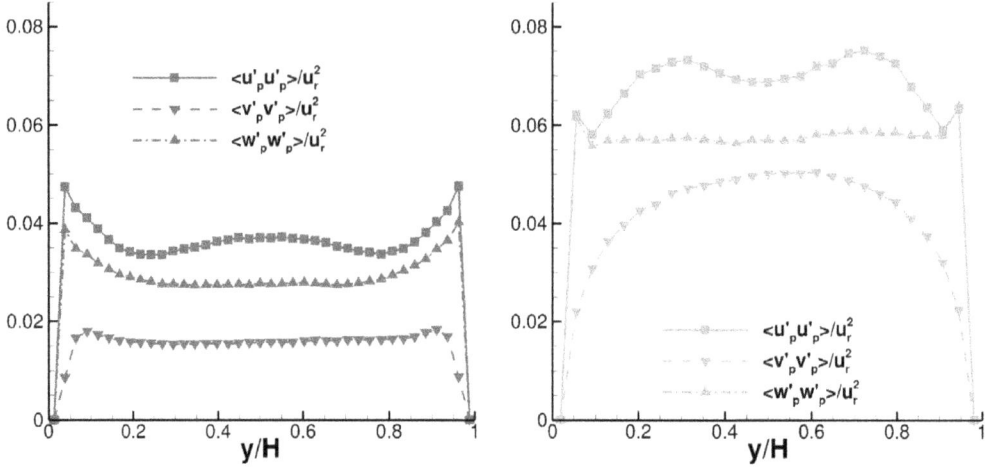

Figure 5.15: Fluctuations of bubbles velocity components when normalized with the bubbles relative velocity for the *SmMany* case (left) and for the *LaMany* case (right).

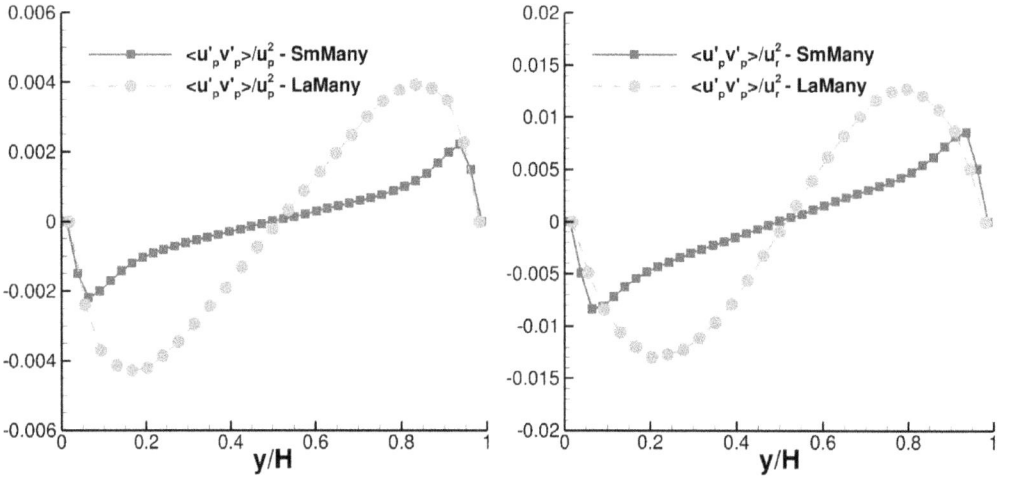

Figure 5.16: Cross-component of bubble velocity fluctuations $\langle u_p' v_p' \rangle$ when normalized with the bubbles rise velocity (left) and with the relative velocity (right).

the side of the bubble that experiences the lower relative velocity. This holds for the small as well as for the large bubbles considered here. In the present channel flow configuration, hence, the lift force moves the bubbles toward the center region, but different shapes of the void fraction distribution are found for the small and for the large bubbles. It may be recalled, that for fixed bubbles the value of the lift coefficient is not so different for the two bubble sizes i.e. Reynolds numbers: $C_y = -0.058$ (steady wake) for the small one and $C_y = -0.071 \pm 0.012$ (oscillatory wake) for the large one. Can this difference explain the different shape observed of the void fraction distribution for small and large bubbles? For a fixed sphere, the values of the lift coefficient were evaluated for different Re_p but at the same dimensionless shear rate. In the channel, in contrast, small and large bubbles experience different shear rates, which can be quantified as

$$\alpha_c(y) = \frac{\Delta \langle u \rangle (y)}{\Delta y}\bigg|_{\Delta y = d_p} \frac{r_p}{u_r(y)} \,. \tag{5.1}$$

For larger bubbles the value of α_c is up to twice as high as the one of small bubbles as portrayed in Fig. 5.17, except in the a narrow region very close to the wall (where large bubbles are almost absent). The lift force acting on large bubbles is, hence, larger than the one on small bubbles, due to the higher Reynolds number *and* due to the higher shear stress. For large bubbles in the channel, then, the influence of the lift force overcomes the influence of the turbophoresis effect (which would move the bubbles toward the wall) and eventually determines the observed peak of $\langle \phi \rangle / \phi_{tot}$ for large bubbles in the core region.

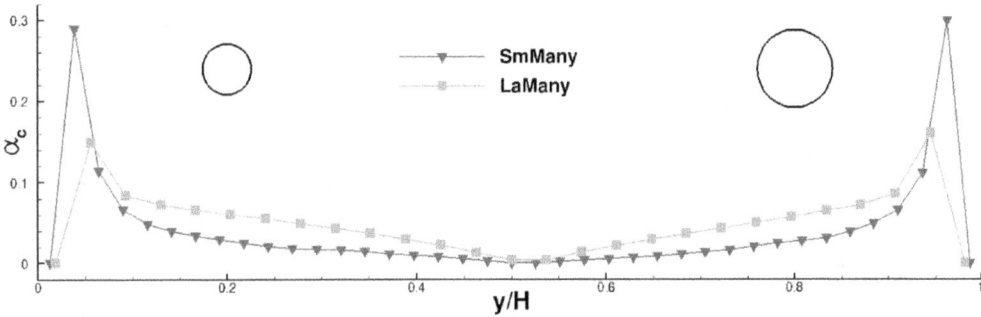

Figure 5.17: Shear rate in channel flow for simulations *SmMany* and *LaMany*. Bubble size represented by circles in the upper part: Left, Small bubble. Right: Large bubble.

5.2.4 Pair correlation functions

The r-PCF of the large bubbles is presented in Fig. 5.18. It shows a similar behavior as the one of small bubbles of the *SmMany* case. The most probable pair distance has also the same values, namely $4r_p$, as for *SmMany*. When compared with the r-PCF of randomly distributed bubbles (RDB), three zones are found. In the first zone, $2r_p \lesssim r \lesssim 3r_p$, the r-PCF is smaller than the random distribution: Bubbles repel each other for very small distances, as investigated by Kim et al. (1993) in the case of fixed sphere and reported in Sec. 3.3.5. For $3r_p \lesssim r \lesssim 15r_p$ bubble pairs are more probable than for the RDB, suggesting this length scale to be the clustering length scale for the present channel flow configuration. For larger distances, the r-PCF is slightly below the random case so that no a significant number of clusters of the size is not expected. The three regions observed are also found for the swarm of small bubbles. Note that, for the large bubbles, half of the channel width $H/2$ corresponds to $13.2r_p$.

The angular pair correlation function (a-PCF) of the bubbles in the *LaMany* case is presented in Fig. 5.19. For small distances bubbles align mainly horizontally, due to the lateral low-pressure zone around the bubbles which induces an attraction of the two bubbles as described in Sec. 3.3.5, by means of what is generally referred to as Bernoulli effect. This feature is

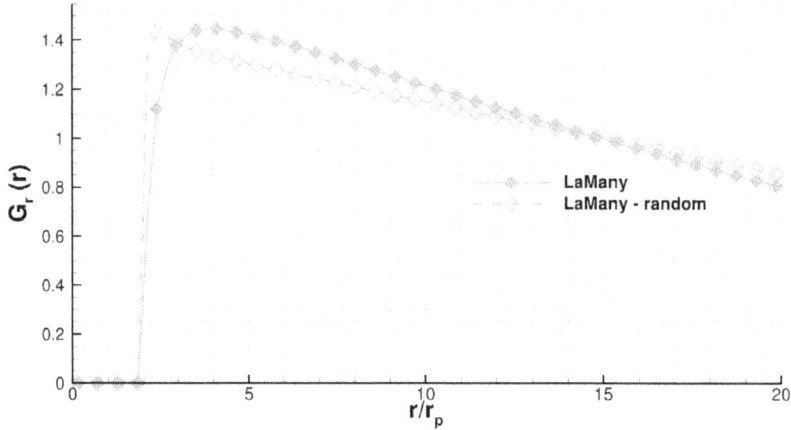

Figure 5.18: Radial PCF for *LaMany* and for randomly distributed bubbles.

reduced with increasing distance and for distances larger than $10d_p$ the a-PCF resembles the one of RDB (not shown here). A closer look at the comparison of the a-PCF for small and large bubbles, provided in Fig. 5.20, allows following considerations. The frequency of the horizontal alignment is reduced for large bubbles with respect to the small bubbles. At first, this may seem unexpected, since not only the bubble Reynolds number but also the relative velocity is larger for the larger bubbles, so that the Bernoulli effect should be stronger. Indeed, the reduced horizontal alignment is due to the more complex bubble dynamics, which increases the bubble agitation and is reflected in more irregular bubble paths thus reducing the small-scale interaction of bubbles. A second observation when comparing the two results in Fig. 5.20 is that the minimum of the a-PCF is located at $\phi_g = 0$ and $\phi_g = 180°$ for the larger bubbles and at $\phi_g = 35°$ and $\phi_g = 145°$ for the *SmMany* case. Hence, there is a slightly stronger tendency of bubbles to follow leading bubbles in their wake when going from *LaMany* to *SmMany*. It is conjectured this effect is also due to stronger deviations of the bubble path from the vertical direction of the larger bubbles. These observations suggest that the different trajectories exhibited by small and large bubbles in quiescent fluid (see Sec. 3.2.1 and 5.1.1) hold also, to a certain extent, in the turbulent channel flow investigated here.

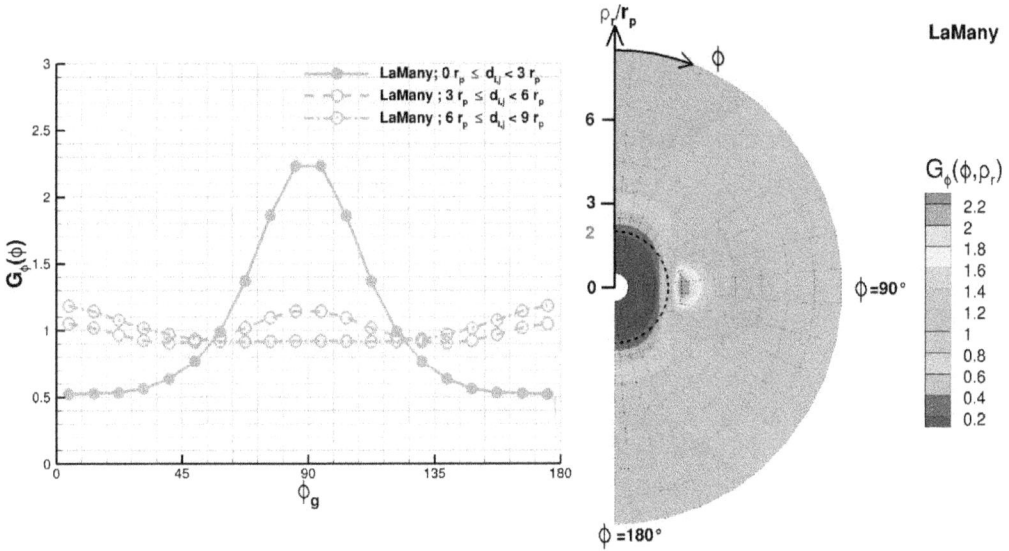

Figure 5.19: Angular PCF for *LaMany*. Left: Canonical representation, different curves for $d_{i,j}/r_p \in [0;3[$, $d_{i,j}/r_p \in [3;6[$ and $d_{i,j}/r_p \in [6;9[$. For this picture $r_2 - r_1 = 3\,r_p$. Right: Two-dimensional representation, with $\rho_r = (r_2 - r_1)/2$ and $r_2 - r_1 = r_p$.

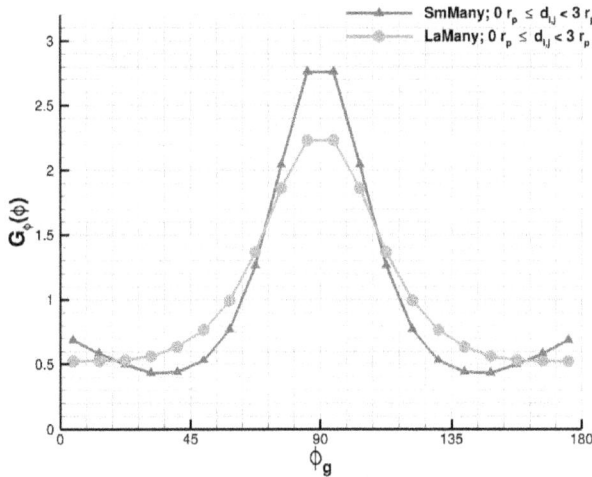

Figure 5.20: Comparison of the Angular PCF for *SmMany* and *LaMany*, small distances $d_{i,j}/r_p \in [0;3[$.

5.3   Results for the bidisperse swarm

This section discusses the case *BiDisp* with the bubble swarm composed of bubbles of the two size classes previously investigated in the monodisperse cases *SmMany* and *LaMa*

5.3.1   Fluid phase: Flow structures and correlation functions

A qualitative impression of the flow field in the channel laden with the bidisperse swarm is provided in Fig. 5.4, right, where instantaneous three-dimensional fluctuations according to (3.5) are shown. With the chosen value of $u'_{3d}/U_b = -0.5$, a velocity structure of negative fluctuations is observed, spanning the whole channel in the streamwise direction. A comparison can be made with two swarms previously investigated, *LaMany* and *SmMany*, portrayed in Fig. 5.4, left, and in Fig. 3.7, respectively, where the same threshold value for the representation of such structure is employed[1]. The flow structures induced by the bidisperse swarm is somewhat longer and wider with respect to the one in the *SmMany* case, but not so extended, in streamwise and spanwise direction, as the one in the *LaMany* case. The influence of the bidisperse swarm on the fluid instantaneous fluctuating flow, hence, seems to be between the influence of the two monodisperse swarms of the two bubble sizes, as expected.

This feature is confirmed by the two-point correlation function of velocity fluctuations in the streamwise direction shown in Fig. 5.5 and 5.6 above. The correlation for the bidisperse case lies always between the the two curves of the monodisperse swarms. In both regions, near the wall and in the center, the small scale behavior is more similar to the one of the *SmMany* case (up to $\Delta x/H = 0.8$ in the wall-region and up to $\Delta x/H = 0.3$ in the center region), while the large-scale behavior is more similar to the one of the *LaMany* case. This suggests that the large-scale flow features are more influenced by the large bubbles than by the small ones.

Two-dimensional fluctuations induced by the bidisperse swarm in the yz-plane according to (3.7) are shown in Fig. 5.21 and two regions can be observed. The picture resembles the corresponding ones of the monodisperse dense swarms, in Fig. 3.11 and 5.7. The flow is dominated by such large structures and near-wall structures are absent. Nevertheless, the fluctuation magnitudes are more similar to the ones in the *LaMany* case, and this is related to the higher level of turbulence produced by the large bubbles.

The two-point correlation function of the velocity fluctuations in spanwise direction is portrayed in Fig. 5.9 and Fig. 5.10 for the near-wall region and for the channel center, respectively. In the near-wall region the correlation is between the curves of the monodisperse swarms up to $\Delta z/H \approx 0.35$. Then, up to $\Delta z/H \approx 0.8$, the correlation is higher than the other functions and it is not between the two other curves. This still results from the intermediate situation, with the correlation length substantially increased compared to *SmMany* but $R_{uu}(\Delta z)$ still levering off beyond $\Delta z/H \approx 0.8H$. Still, the slope is not as large as for *LaMany*, where the stronger spanwise interaction results in a negative correlation for $\Delta z/H > 0.6$. This strong interaction prevail in the center region for all the cases with only little differences of R_{uu} for $\Delta z/H > 0.4$. Furthermore, the correlation of *BiDisp* is between the ones of the monodisperse swarms, with a slightly larger resemblance to the *SmMany* case in the small scales, where small bubbles have a larger influence on turbulence production

---

[1]For the three cases *LaMany* and *SmMany* and *BiDisp* the global void fraction is the same, as reported in Tab. 3.2

and, thus, on the small-scale reduction of the correlation.

In conclusion, the influence of the present bidisperse swarm on the fluid velocity structures is similar to the one of monodisperse swarms with the same total void fraction, with small bubbles generating a larger imprint on small-scale behavior and large bubbles a more pronounced effect on the large-scale flow features.

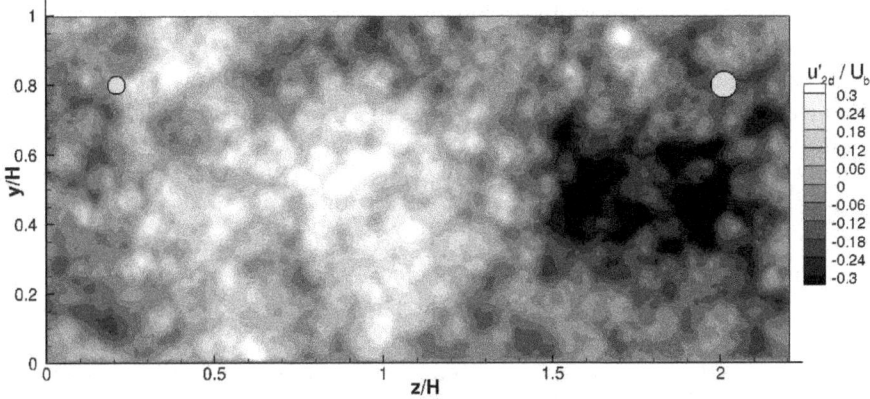

Figure 5.21: Instantaneous two-dimensional velocity fluctuations u'_{2d} according to (3.7) for simulation *BiDisp*. The bubble sizes are represented by two circles at $y/H = 0.8$, $z/H = 0.2$ (small bubble) and at $y/H = 0.8$, $z/H = 2$. (large bubble).

5.3.2 One-point statistics of the carrier phase

As collected in Table 3.2, the time-averaged shear Reynolds number in the *BiDisp* case is $Re_\tau = 194.1$. This lies between the ones of the monodisperse swarms and its fluctuations around the mean value are not the pronounced as in the *LaMany* case. Figure 5.11 displays the mean velocity profile in the bidisperse swarm. As expected, it lies between the velocity profiles of the monodisperse swarms. When represented in wall units, the velocity profile exhibits some similarities with the *LaMany* case, thought the value is always lower and it exhibits a small logarithmic region over the range $y^+ \approx 15...50$. Obviously the profile is dominated by the input of the larger bubbles as the slope of $\langle u \rangle^+$ resembles the one for *LaMany*.

The non-vanishing components of the Reynolds stress tensor are portrayed in Fig. 5.12 and compared with the corresponding components of the two monodisperse swarms. All profiles for the *BiDisp* case lie between the curves of *SmMany* and *LaMany* and have a shape similar to the case *LaMany*, just less pronounced. Only for the streamwise component $\langle u'u' \rangle$ very near the wall, $y_w/H < 0.055 \approx d_{p,SM}$, the fluctuations of the *BiDisp* case are lower than in the *SmMany* case, due the larger amount of small bubbles rising in the near-wall region. In this intermediate case the wall-normal and the spanwise stress component are uniform over a large portion of the channel.

5.3.3   One-point statistics of the disperse phase

For the statistical analysis of bubble quantities, the width of the wall-normal bins was chosen equal to the radius of the small bubble, i.e. $I_w = r_{p,SM}$. The choice of the storage frequency of bubble data was again based on the mean bubble velocity, that is the average between large and small bubble velocity.

Figure 5.22 shows the bubble statistics for the *BiDisp* case when averaging is performed over all bubbles in the swarm. The fact that both constant void fraction and constant relative velocity prevail over a large part of the channel may be at the origin of the relative constant values of the Reynolds stress components $\langle v'v' \rangle$ and $\langle w'w' \rangle$ over the same region.

When analyzing small and large bubbles separately, the different behavior observed in the monodisperse swarms is noticed again. The void fraction of small bubbles presents a pronounced peak at the wall while large bubbles rise prevalently in the center region, as depicted in Fig. 5.23, top. When these profiles are compared with the ones of the monodisperse swarms, it is observed that both the tendency of small bubble to rise in the near-wall regions and the tendency of large bubbles to rise in the center region are enhanced (cf. Fig. 3.19, Sec. 3.3.3, and Fig. 5.13, Sec. 5.2.3). This enhancement seems to be somewhat stronger for the small bubbles, while the distribution of the large bubbles is only slightly modified with respect with the *LaMany* case. The dominant mechanisms that determine such behavior, i.e. the turbophoresis effect for the small bubbles and shear-induced lift force for the large bubbles, are hence conjectured to be dominant also in a bidisperse swarm. The tendency of small bubbles to rise prevalently in the near-wall region is also the reason of the enhancement of the wall-shear stress with respect to the *Unladen* case, as collected in Tab. 3.3. As for the *SmMany* case, the higher momentum exchange in the wall-normal direction is due to the interaction between wall turbulence and the wakes of small bubbles rising close to the wall, inducing regions with large velocity gradients. The tendency of large bubble to rise in the center region is also the reason of the modified velocity profile $\langle u \rangle /U_b$ portrayed in Fig. 5.11, since faster, larger bubble drag the fluid yielding higher values of $\langle u \rangle$. Regarding the bubble velocity, as expected large bubbles rise faster than small bubbles and for both bubble classes the relative velocity is practically uniform across the channel.

Bubble velocity fluctuations are also evaluated for the two size classes separately, according

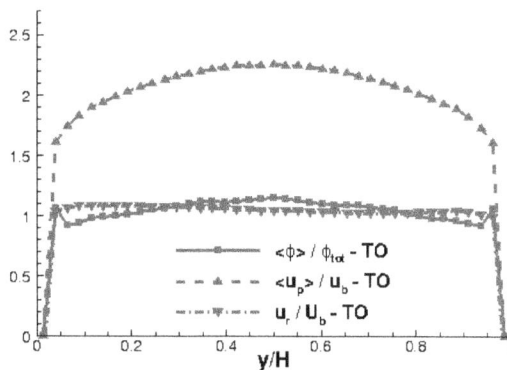

Figure 5.22: Bubble-related statistics for case the *BiDisp* when average is performed over both bubbles classes: Void fraction distribution, rise velocity and relative velocity.

to

$$u'_{p,SM}(y) = u_{p,SM}(y) - \langle u_{p,SM} \rangle_{xzt}(y) , \qquad (5.2)$$

with analogous definitions for the large bubbles and for the other velocity components. Profiles of bubble velocity fluctuations are shown in Fig. 5.23, bottom. It can be observed that bubble fluctuations of all velocity components present comparable magnitudes for small and large bubbles. The shapes of the streamwise fluctuation profile are slightly different in the wall-regions: For small bubbles, $\langle u'_{p,SM} u'_{p,SM} \rangle$ increases smoothly with the wall distance, while for the large bubbles it presents a maximum for the at the wall nearest possible position. This was also observed in Sec. 5.2.3, where the dense swarm of large bubbles *LaMany* was analyzed. When compared with the bubble fluctuations of the monodisperse swarms, the following observations can be made. The fluctuation levels of the small bubbles in the bidisperse swarm are higher than the corresponding ones in the monodisperse swarm with small bubbles (see Fig. 3.19). Note that all fluctuations are scaled with bulk unit, i.e. with U_b^2. This is due to the presence of the large bubbles in the bidisperse swarm, which enhance the turbulence level of the flow yielding larger fluctuations of the small bubbles. This was observed also with the components of the Reynolds stress tensor, presented in the in Fig. 5.12, that hint at an overall higher turbulence level for the *BiDisp* case than in the *SmMany* case. Another observation is that the shape of the fluctuations of the small bubbles are similar to the fluid fluctuations (Fig. 5.12) suggesting a stronger dependence between fluid and small bubbles. The large bubbles in the *BiDisp* case, instead, present fluctuation profiles lower than the corresponding ones in the monodisperse swarm with large bubbles, case *LaMany*. This is also related to the smaller number of large bubbles in this swarm inducing lower turbulence of the flow.

5.3.4 Pair correlation functions

The r-PCF of the bubbles in the bidisperse swarm, presented in Fig. 5.24, is evaluated taking into account all bubbles. All distances involved, as in the a-PCF analysis below, are scaled with the radius of the small bubbles. The most probable pair distance is around $5.2 r_{p,SM}$. When comparing the r-PCF with the one of bidisperse RDB, two regions are observed. In the first region, for $2 \lesssim r/r_{p,SM} \lesssim 5$, pairs in the swarm are less probable than in the random case. This is due to the small-scale repulsion of bubbles, as discussed in Sec. 3.3.5. In the second region, for $r/r_{p,SM} > 5$, pairs are as probable as in the random case. This means that, in contrast with the monodisperse swarms, the tendency of the bubbles to rise pairwise is lower in the bidisperse swarm. This is not unexpected as the rise velocity of small and large bubbles is different. The a-PCF of bubbles in the bidisperse swarm is shown in Fig. 5.25, where all bubbles are considered and the frequency of pair alignment in different shells is investigated. For small distances, the preferential horizontal alignment of bubbles is observed also in the bidisperse swarm, as in the monodisperse ones.

When compared with the a-PCF of small bubbles in the *SmMany* case (Fig. 5.20), the frequency of the horizontal alignment of bubbles is reduced. This is due to the higher turbulence level induced in the flow by the bidisperse swarm, as already discussed in the previous section and confirms the somehow more irregular dynamics of small bubbles in the *BiDisp* case. Analogously, the horizontal alignment of large bubbles is increased with respect to the monodisperse swarms, case *LaMany*. This is due to the lower turbulence level and hints at a more regular dynamics of large bubbles in the bidisperse swarm, confirmed also by the local maxima at $\phi_g = 0°$ and $\phi_g = 180°$ for large-large bubble pairs (a position

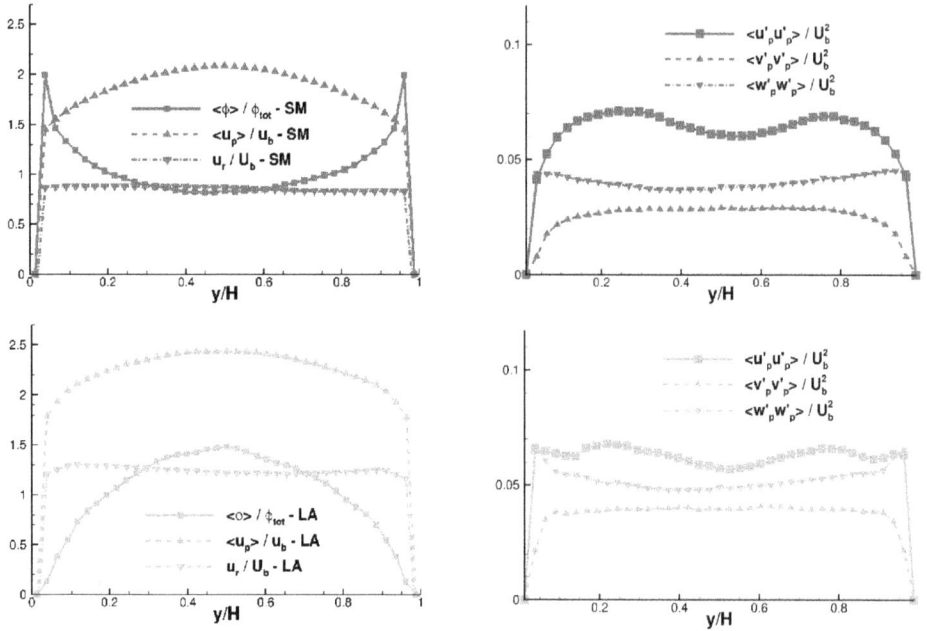

Figure 5.23: Bubble-related statistics for case *BiDisp* for separate bubble classes. Top: Void fraction distribution, rise velocity and relative velocity for small bubbles (left) and for large bubbles (right). Bottom: Bubble velocity fluctuations for small bubbles (left) and for large bubbles (right).

of unstable equilibrium, as reported in Sec. 3.3.5 when the a-PCF of small bubbles in the monodisperse swarm was considered).

Interesting information can be obtained when only mixed pairs are considered, i.e. pairs constituted by a small and a large bubble. For this analysis, the origin of the coordinate system is placed at the center of the small bubble. Results of the a-PCF of mixed pairs for different distances are shown in Fig. 5.26. For small distances, the most probable alignment is found for $\phi_g \approx 60°$ and this feature is highly reduced when the pair distance increases, as represented also in the two-dimensional representation of the a-PCF. The sudden drop of the frequency of the 60°-alignment suggests that this a small-scale feature. A preferential oblique alignment was also observed by Göz and Sommerfeld (2004) for a bidisperse swarm rising in quiescent fluid, where the void fraction was 6% and the volume ratio was 2^2. These authors observed that a small bubble stays "sideways to obliquely behind a large bubble" and explained this feature with the different rise velocity. The oblique alignment, as suggested by the present results, holds also for bubbles in turbulent flows and deserves a deeper analysis. The instantaneous pressure field around a mixed pair aligned at $\phi_g \approx 60°$ in a vertical plane cutting the bubble centers is displayed in Fig. 5.27 and the lateral low pressure zone around both bubbles can be observed. The pressure field around the large bubble spreads out to a longer extent, as expected. As for the interaction of bubbles of equal size, the low-pressure region is responsible for the attraction of bubbles at intermediate distance. In this case the attraction that the large bubble excerts on the small one is larger than the attraction

---

[2]Here, the volume ratio is 3.16

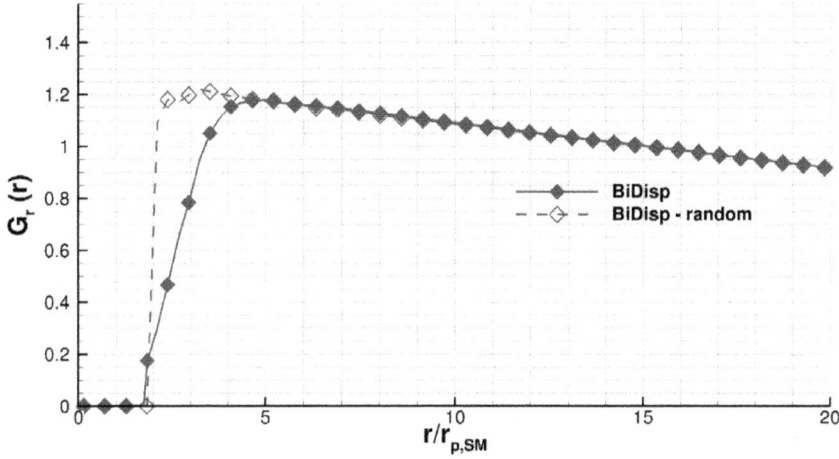

Figure 5.24: Radial PCF for *BiDisp* and randomly distibuted bubbles, considering all bubbles in the swarm. The distance r is scaled with the radius of the small bubbles $r_{p,SM}$.

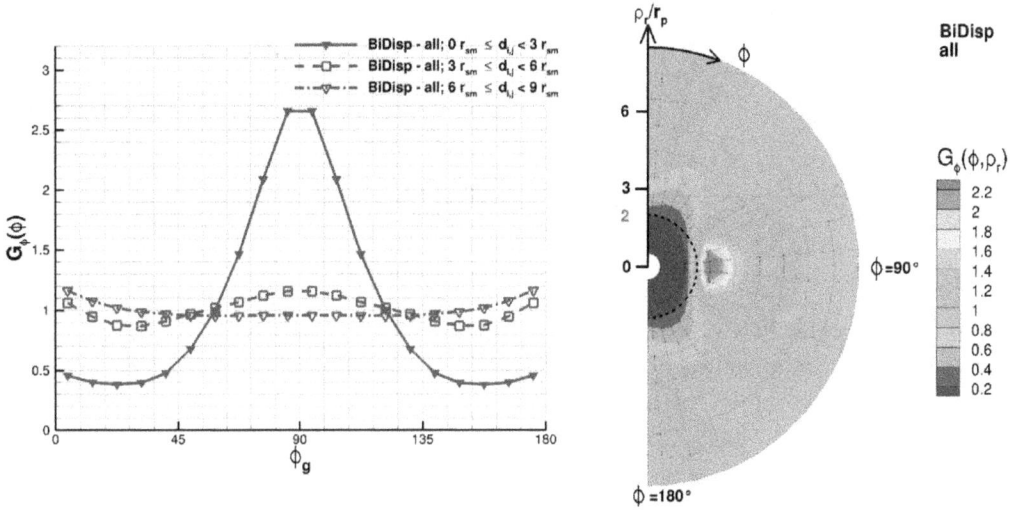

Figure 5.25: Angular PCF of bubbles in the *BiDisp* case without distinction of bubble size. Right: Canonical representation for different shells. Left: Two-dimensional representation with $\rho_r = (r_2 - r_1)/2$ and $r_2 - r_1 = r_p$.

of the small one on the large one. Since the large bubble rises faster than the small one, this mechanism does not result in a horizontal alignment, but rather in an oblique one, as mentioned above. This feature is further discussed by means of the time history of selected quantities of the same mixed pair depicted in Fig. 5.27. Figure 5.28 shows the relative distance of the bubbles $d_{i,j}$, its projection on the horizontal plane $d_{y,z}$, the angle between the vertical direction, and the segment connecting the bubble centers $\phi_{i,j}$ and the streamwise instantaneous velocity difference Δu_p between the large bubble and the small one. The following interaction can be deduced from the observation of such quantities. The large bubble approaches the small bubble from below, eventually reaching it because of the higher velocity. When the distance is about $3r_{p,SM}$ ($\Delta t/T_b \approx 0.53$ in Fig. 5.28) the bubbles start to interact: The small bubble moves toward the large one and it also accelerated by the latter, as shown in Fig. 5.29, where bubble rise velocities are displayed. It is interesting to observe that the acceleration of the small bubbles takes place for $d_{i,j}/r_{p,SM} \leq 3$, both before and after the instant of smallest distance. The large bubble, instead, is slightly decelerated by the small one, even if such deceleration, caused by a local higher stress on the large bubble, is lower than the acceleration experienced by the small one.

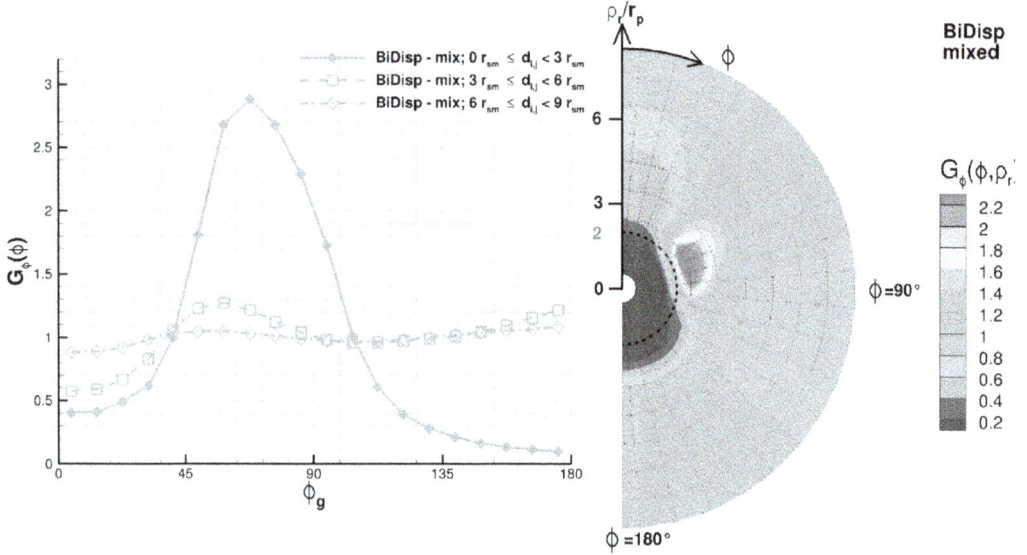

Figure 5.26: Angular PCF of bubbles in the *BiDisp* case, when considering only mixed pairs. Right: Canonical representation for different shells. Left: Two-dimensional representation with $\rho_r = (r_2 - r_1)/2$ and $r_2 - r_1 = r_p$.

In conclusion, the analysis of a monodisperse swarm of larger bubbles, case *LaMany*, of a bidisperse swarm, case *BiDisp* and the comparison with the swarm of small bubbles, case *SmMany*, can be summarized as follows. The swarm of larger bubbles generates elongated flow structures in the streamwise directions that are not observed in the single-phase flow. These structures, visualized by fluid velocity fluctuations, are longer than the corresponding ones generated by a swarm of small bubbles with the same total void fraction. A bidisperse swarms induces flow structures whose size is between the ones of the monodisperse swarms. The elongated structures are also observed when by means of two-dimensional fluctuation regions are represented, when average is performed over the streamwise direction. Two regions of alternating fluctuation sign, spanning almost the whole channel width H, are found in the yz-plane, and the picture is similar for the three swarms investigated. Channel flow statistics are strongly modified by the presence of bubbles. For the parameter range investigated here, bubbles enhance the flow turbulence, since the production of turbulence is dominated by the so-called pseudo-turbulence. This augmentation, quantified by the non-vanishing component of the Reynolds stress tensor, is larger in the case of larger bubbles. This is due to the

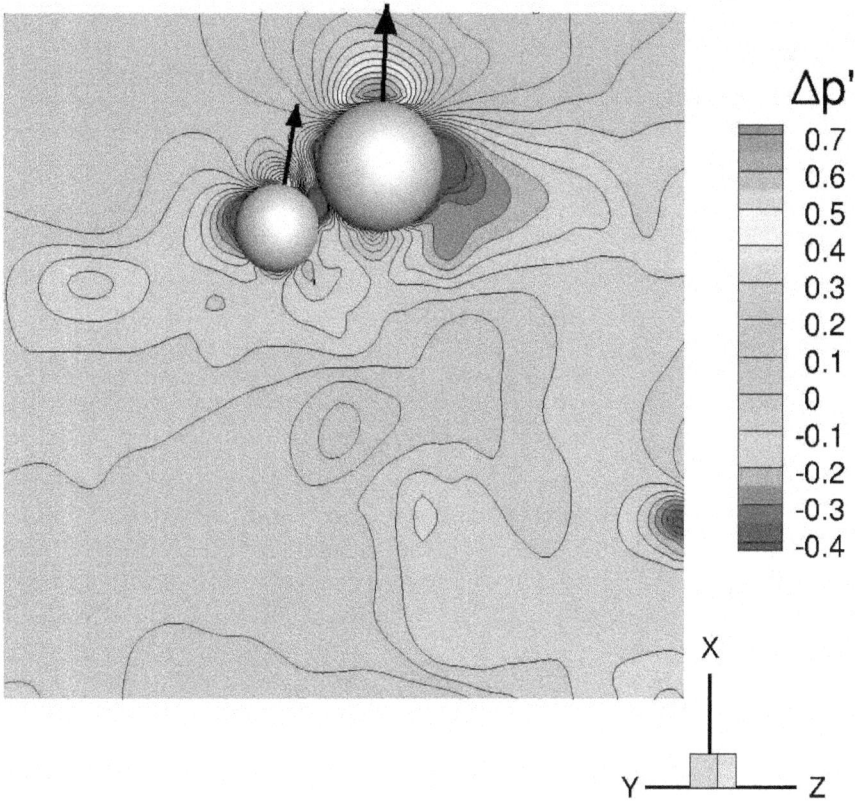

Figure 5.27: Fluid pressure field around a mixed pair of bubbles in a vertical plane cutting the two bubble centers.

dynamics of larger bubbles, which present a more complex paths and strongly modify the flow. In the present configuration, small bubbles rise preferentially in the near-wall region and large bubbles in the center region. In the first case, the turbophoresis effects plays a major role, while in the second case the lift force induced by the mean fluid velocity profile dominates. The different dynamics of bubbles as function of the Reynolds number observed in quiescent fluid was also found in the channel flow. The different behavior was quantified, among others, by means of the bubble velocity fluctuations: As expected, these are larger for larger bubbles, even when scaled with the rise and with the relative velocity. In the bidisperse swarm, an interesting feature was observed: With respect to the monodisperse swarms, fluctuations of large bubbles are reduced and fluctuations of the small are increased. Large bubbles, in fact, enhance the fluid turbulence which induces higher fluctuations of the small bubbles. Bubbles in the three denser swarms align mainly horizontally due to the later low-pressure zones around rising bubbles. In the swarm consisting of large bubbles the vertical alignment is the less probable, due to the irregular bubble dynamics: It occurs rarely that a trailing bubble can follow the path of a leading one. In the bidisperse swarm, the most probable alignment of "mixed"-pairs is an oblique one, where the segment connecting

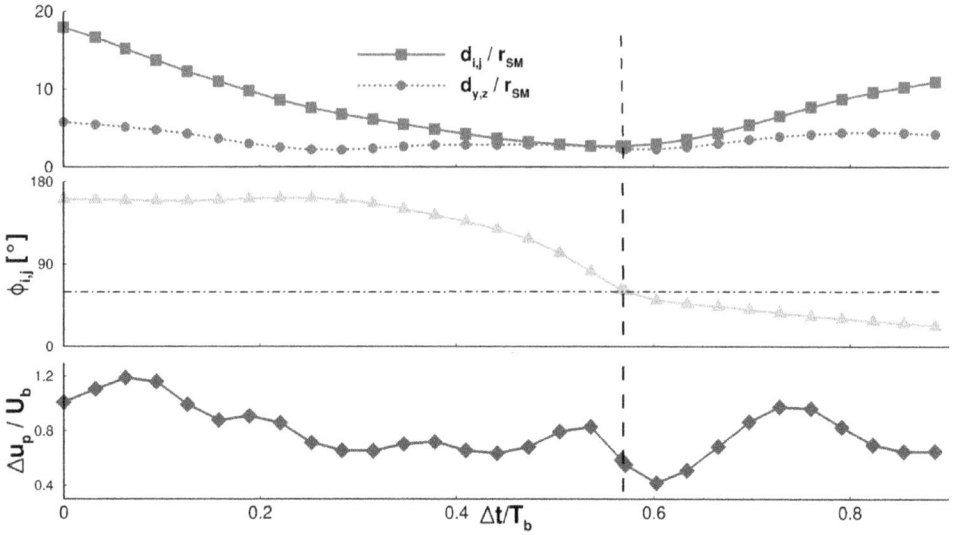

Figure 5.28: Time history of bubble quantities for a selected mixed pair. Top: Distance between bubble centres and projection of the distance on the yz-plane. Middle: angle between bubble centers. Bottom: bubble velocity difference $\Delta u_p = u_{p,LA} - u_{p,SM}$. The vertical dashed line represents the time instant of Fig. 5.27, the horizontal dash-dotted line indicates $\phi_g = 60°$.

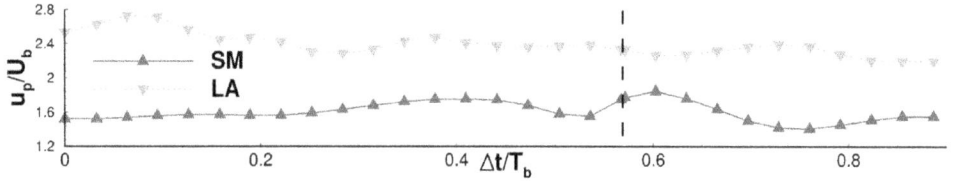

Figure 5.29: Time history of bubble rise velocity. The vertical dashed line represents the time instant of Fig. 5.27.

the bubble centers presents an angle of around 60°with respect to the vertical line. This feature is due to the low-pressure zones around each bubble and due to the different rise velocity.

6 Influence of the flow direction

Downward bubbly flows occur, when the direction of the fluid velocity is opposed to the rise direction of the bubbles. In the past such configurations have not gained the same attention and research effort as the upward configuration, although this type of flow can be found in many industrial applications, such as chemical reactors and drill strings. Many differences arise between upward and downward flows and one important issue is the transfer of energy between the fluid and the bubbles. In upward flows the buoyancy force is directed in the same direction as the force induced on the bubbles by the fluid and this superimposition yields complex interaction, as reported in the previous sections. In downward flows, instead, buoyancy acts against the force induced by the fluid and is needed to overcome the motion that the fluid induces on the bubbles.

Wang et al. (1987) experimentally investigated both the upward and downward flow of small bubbles in the disperse regime. For downward configurations, these authors reported a core peaking of the void fraction distribution, opposed to wall peaking occurring in upward configurations. Sun et al. (2004) experimentally measured streamwise fluid velocity and fluctuations for downward flow in dilute and slug flow regimes. For the disperse regime, a flatter velocity profile was observed in the core region due to the bubble presence. Turbulence enhancement was reported for the investigated regime, but unfortunately not much information was provided. Kashinsky et al. (2008) performed experimental investigations to address the influence of bubbles on the structures of the downward bubbly flow in a pipe and observed that even at low void fractions ($\phi_{tot} < 5\%$) the flow statistics are strongly modified by the presence of bubbles. Lu et al. (2006) and Lu and Tryggvason (2007, 2008) performed DNS of downward flows for both laminar and turbulent flows analyzing the influence of the bubbly size and deformability. These authors described similar phenomena, i.e. that nearly spherical bubbles moved toward the channel center due to the shear-induced lift force. It was shown that, in the downward case, the flow is always in hydrostatic equilibrium between the imposed pressure gradient, the shear of the flow and the buoyancy of the mixture. Recently, Bhagwat and Ghajar (2012) collected a multitude of experimental results in both upward and downward pipe flows and analyzed the influence of the flow direction on the different flow regimes, e.g. disperse bubbly flow, slug flow, etc. For the disperse regime, these author observed an increase of the bubble number and a decrease of the bubble size with increasing flow rate. This feature is induced by the work done by the fluid on the bubbles which "disintegrate" the bubbles into smaller ones. Such considerations are based on the visual inspection of experimental data but were not quantified. Lelouvetel et al. (2011, 2014) performed experimental investigations of both upward and downward flows to investigate the impact of bubbles on the turbulent structures and on the transfer of energy between the phases at different flow scales. These authors reported turbulence reduction for upward flow and turbulence enhancement for downward flow.

To address the dynamics of bubbles in a downward configuration, a simulation was performed where all parameters are set equal to the ones in the denser swarm of small bubbles (case *SmMany* in Chap. 3) except for the direction of the fluid flow. Results of this simulation, labeled *SmManyDo*, are reported in this section and compared with the ones of the *SmMany* case to investigate the role of the flow direction for the chosen parameter range.

6.1 Fluid phase: Flow structures and correlation functions

A first impression of the instantaneous flow field is provided in Fig. 6.1, where three-dimensional velocity fluctuations according to (3.5) are represented. The flow seems more complex than in the *SmMany* case (see Fig. 3.7, right). One large structure is observed which spans almost half of the channel in the streamwise direction when portrayed with the chosen threshold value. Smaller structures are also present and are somehow connected to the large structure. A region of negative fluctuations close to the wall and oriented mainly in the spanwise direction is observed at $y/H \approx 1$ and $0.5 \lesssim x/H \lesssim 1$ (between the first and the second horizontal plane from the bottom). Such a large wall-bounded structure was not observed in the instantaneous flow visualizations of any swarm in upward configuration. The two-point correlation function of the velocity fluctuations in the streamwise direction according to (3.6) is portrayed in Fig. 6.2. In the near-wall region the small-scale reduction of the correlation due to the bubbles with respect to the unladen case is observed. This feature was also reported for the *SmMany* case. The negative values of R_{uu} for large streamwise distances, instead, suggests the presence of structures of alternating fluctuation sign in the streamwise direction. This feature is confirmed by the two-point correlation function in the channel center which also presents negative values for large distances as for the upward configuration. The overall impression is, hence, that large structures of alternating fluctuation sign span the whole channel in the streamwise direction.

The two-dimensional velocity fluctuations according to (3.7) for the downward configuration are portrayed in Fig. 6.3. There is one region of positive fluctuations for $z/H \gtrsim 1.2$ and two regions of negative fluctuations, a larger one centered around $(y/H, z/H) = (0.7, 0.4)$ and a smaller one in the near-wall region, centered around $(y/H, z/H) = (0.1, 0.4)$. Both regions of negative fluctuations, although different in size, are found in the near-wall region. The picture is somehow more complex than in the upward case (cf. Fig. 3.11), where only two large regions of different fluctuation sign were observed.

Such observations are confirmed if one considers the two-point correlation function of the velocity fluctuations in the spanwise direction, $R_{uu}(\Delta z, y)$. As depicted in Fig. 6.4, left, no reduction of the correlation is observed in the near-wall region for small-scale distances with respect to the unladen flow. For large distances the correlation presents a finite positive value and this suggests the existence of structures in the near-wall regions that span the whole channel in the spanwise direction. This feature was observed also in the instantaneous three-dimensional fluctuation field (Fig. 6.1) but not in two-dimensional velocity fluctuations (Fig. 6.3). As shown in Fig. 6.4, right, the correlation in the center region is slightly reduced with respect to the unladen flow for small distances where the production of turbulence induced by the bubbles is noticed. For large distances, R_{uu} is negative which implies the presence of structures of alternating fluctuation sign in the center region, as in the *SmMany* case. Nevertheless, the presence of such structures in a downward bubbly flow should be further addressed and additional simulations (e.g. varying the ratio between buoyancy and drag or

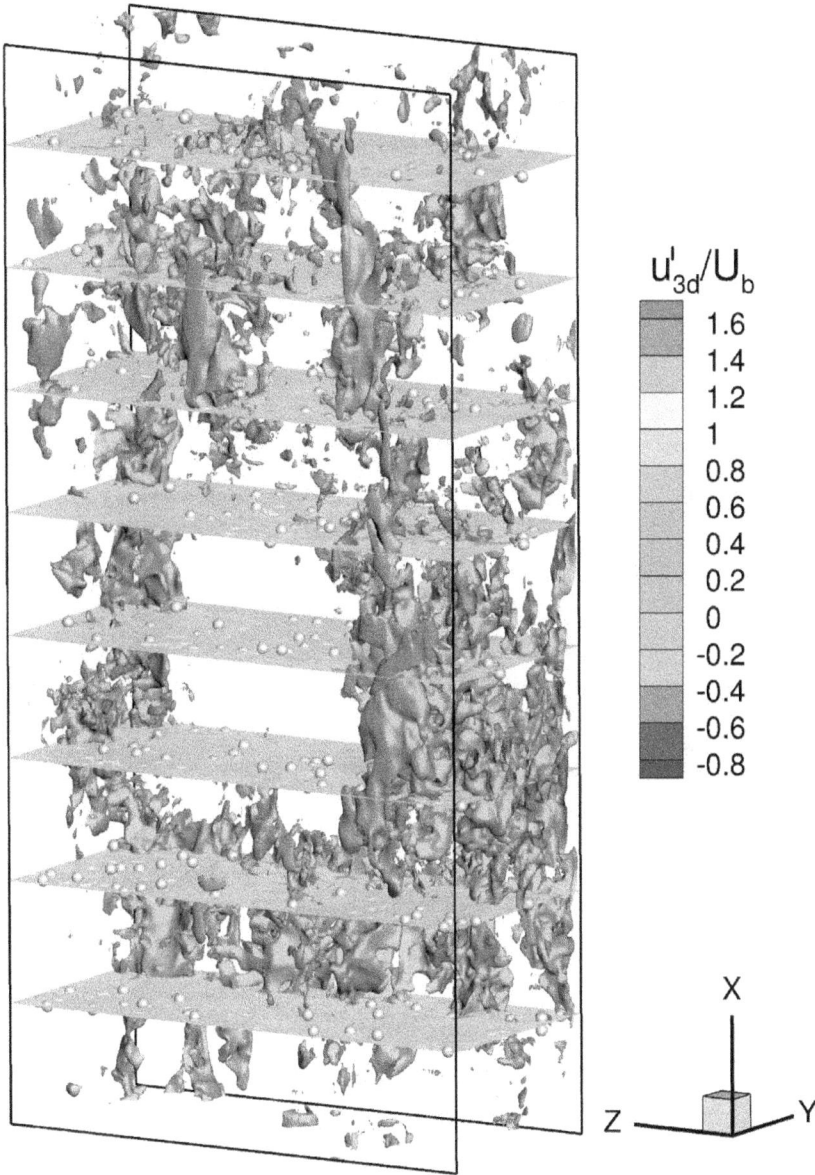

Figure 6.1: Instantaneous three-dimensional velocity fluctuations as defined by (3.5) for simulation *SmManyDo*. The three-dimensional structure is the isosurface $u'_{3d}/U_b = -0.5$ as in Fig. 3.7. The vertical distance between the planes used for contour plots is $0.5H$. Only bubbles cutting such planes are represented.

the bubble size) are required to gain a deeper insight.

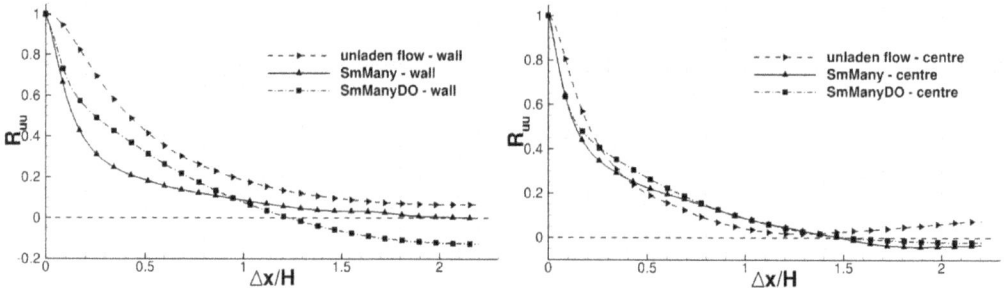

Figure 6.2: Two-point correlation function R_{uu} according to (3.6) in the streamwise direction. Left: Near-wall region, $y/H \approx 0.02$. Right: Channel center.

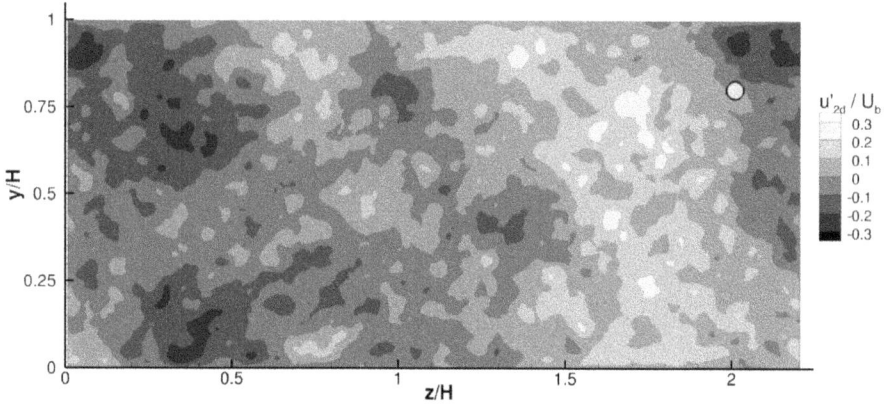

Figure 6.3: Instantaneous two-dimensional velocity fluctuations u'_{2d} according to (3.7) for simulation *SmManyDo*. The bubble size is represented by a circle at $y/H = 0.8$, $z/H = 2$.

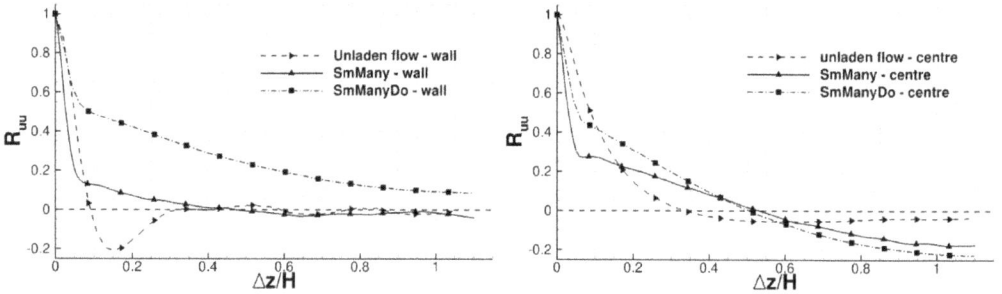

Figure 6.4: Two-point correlation function R_{uu} according to (3.6) in the spanwise direction. Left: Near-wall region, $y/H \approx 0.02$. Right: Channel center.

6.2 One-point statistics of the carrier phase

The averaged fluid velocity in the streamwise direction $\langle u \rangle /U_b$ is portrayed in Fig. 6.5. When compared to the *SmMany* and the *Unladen* case, the profile is steeper at the walls and flatter in the core region. A flatter velocity profile was also observed by Wang et al. (1987), Sun et al. (2004), Kashinsky et al. (2008) for downward flows and this feature is related to the presence of bubbles in the core region as addressed below. The larger velocity gradient at the walls yields a larger shear Reynolds number and its time-averaged value is 263.9, larger by a factor 1.25 and 1.56 compared to the *SmMany* and *Unladen* case, respectively. When normalized in wall units, Fig. 6.5 right, the velocity profile presents lower values due to the higher shear stress velocity and two regions are observed: For $y^+ < 20$, it is very similar to the *SmMany* case, while for $y^+ > 20$ it present an almost logarithmic shape with lower inclination.

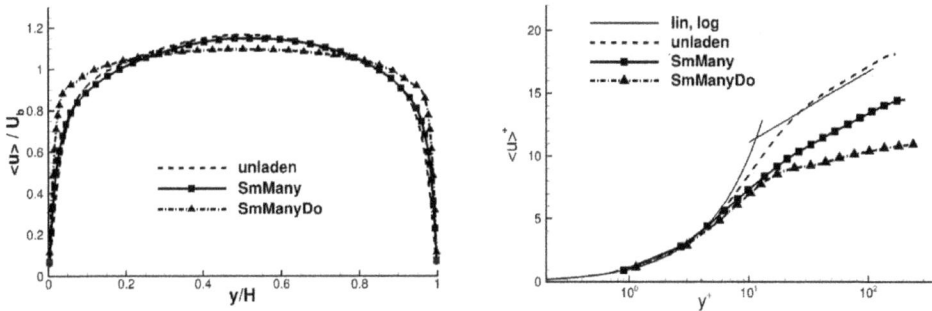

Figure 6.5: Averaged fluid streamwise velocity for *Unladen*, *SmMany* and *SmManyDo* For the latter, the quantity $\langle u \rangle$ is invserted in sign. Left: Scale with bulk units. Right: Scale with wall units (profiles averaged on both channel sides).

The non-vanishing components of the Reynolds stress tensor for the *SmManyDo* case are portrayed in Fig. 6.6 and compared to the corresponding quantities of the *SmMany* and *Unladen* cases. The shape of the streamwise fluctuation profile $\langle u'u' \rangle$ resembles the one of the unladen case, but with much higher values. In contrast with the *SmMany* case, no flat maximum is presented in the core region. The profile of the wall-normal component $\langle v'v' \rangle$ is quite different with respect to the two other cases. Up to $y_w/H = 0.05$ it matches with the profile of the *SmMany* case, but it increases with wall distance up to the center of the channel, where the value is more than three times the one of the *SmMany* case. For the fluctuations $\langle w'w' \rangle$ in the spanwise direction, the shape is similar to the one of the other cases but higher values are observed for this quantity as well. For the turbulent shear stress $\langle u'v' \rangle$, for $y_w \lesssim 0.2$ values are larger as for *Unladen* and smaller as for *SmMany* case, while in the core region the profile collapses with the latter. Therefore it can be stated that in the downward configuration turbulence enhancement is observed when compared both to the unladen channel and with the upward bubbly flow under otherwise the same conditions. An increase of turbulence intensity due to the bubble presence in downward flow was reported also in the literature (Lelouvetel et al., 2011; Sun et al., 2004; Kashinsky et al., 2008) so that the turbulence enhancement observed here matches with experimental investigations. Eventually, the analysis of the components of the Reynolds stress tensor suggests a substantially larger enhancement of the turbulence level in the downward flow when compared to the upward case.

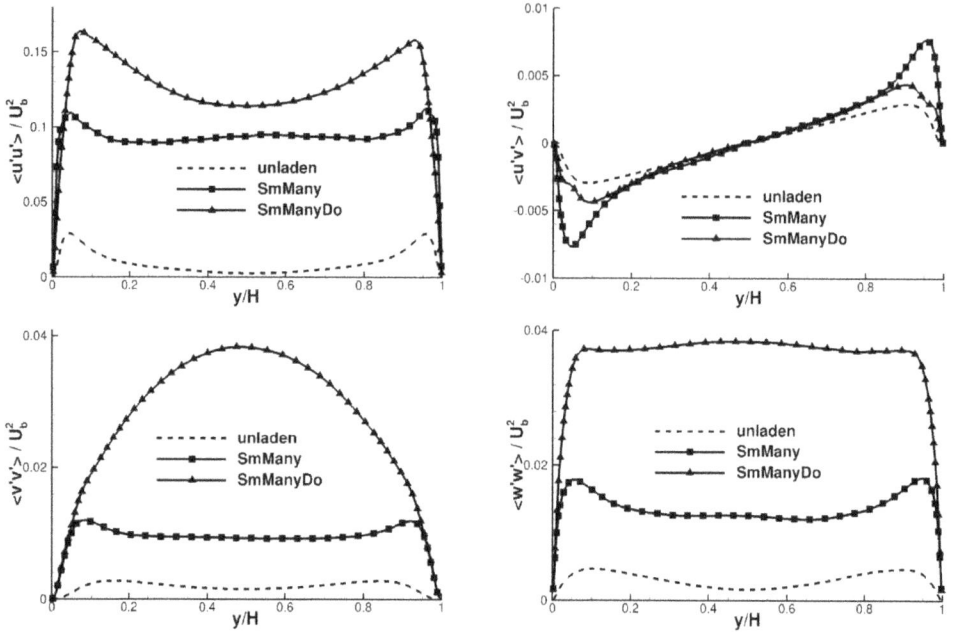

Figure 6.6: Reynolds stress components for *Unladen*, *SmMany* and *SmManyDo*. Top, left: Streamwise fluctuations $\langle u'u' \rangle$; top, right: Turbulent shear stress $\langle u'v' \rangle$ (inverted in sign for *SmManyDo*; bottom, left: Wall-normal fluctuations $\langle v'v' \rangle$; bottom, right: Spanwise fluctuations $\langle w'w' \rangle$.

6.3   One-point statistics of the disperse phase

Bubble-related statistics are portrayed in Fig. 6.7. The void fraction distribution presents two peaks in the near-wall region and is fairly constant in the interval $0.1 < y/H < 0.9$. The position of the peak is slightly further away from the wall than in the *SmMany* case, shown in Fig. 3.19, left. The rise velocity of the bubbles $\langle u_p \rangle$ presents slightly positive values very close to the walls and then decreases smoothly and reaches its negative maximum in the channel center. This means that for the chosen parameter set (buoyancy and flow rate) the bubbles slowly move downward in the channel, except in the proximity of the walls. The relative velocity of bubbles, defined according to (3.16) as in the upward cases, is fairly constant across the channel, slightly lower in the center region. The shape is slightly different when compared to the relative velocity of bubbles in the upward case, as portrayed in Fig. 3.19, left. Nevertheless, the magnitude of u_r is very close to the one of the *SmMany* case. This can be expected since the forces acting on the bubbles (buoyancy and fluid-induced drag) are about the case in this case. Based on the relative velocity and the void fraction distribution, the mean bubble Reynolds number according to (3.17) is 221.3, which is slightly lower than the one of the bubbles in the upward channel.

The small positive value of the bubble velocity close to the wall deserves a few comments. This feature was also observed by Sun et al. (2004) and Lelouvetel et al. (2014) and both research teams supposed this feature to be related to difficult measurements for the regions very close to the wall. The present results show that a small increase of the bubble velocity close to the wall is observed in numerical simulations as well and is, therefore, not related to the lack of accurate measurements. Unfortunately, the velocity of the bubbles close to the wall is not reported in (Lu et al., 2006) and (Lu and Tryggvason, 2008) and a comparison with other DNS data is therefore not possible. What is observed here, is that in a narrow region very close to the wall the buoyancy overcomes the force induced by the low fluid velocity: This yields, in the present case, the slightly positive values of $\langle u_p \rangle$ portrayed in Fig. 6.7.

The profiles of the bubble velocity fluctuations are portrayed in Fig. 6.7, right, and are normalized with the relative velocity u_r. This decision was taken to provide a direct comparison with the bubble velocity fluctuation in the upward direction. The streamwise component $\langle u_p' u_p' \rangle$ presents two maxima close to the walls, at around the same position of the maxima of the void fraction distribution and of the Reynolds stress tensor component $\langle u'u' \rangle$. It then decreases and reaches its minimal value at the channel center. The spanwise component has a similar shape, but the reduction in the center region is much lower, yielding a fairly constant value for $0.1 < y/H < 0.9$. The shape of the wall-normal component, instead, is quite different: It increases smoothly with increasing wall distance and reaches it maximal value in the channel center, where the value of $\langle u_p' u_p' \rangle$ is comparable with the one of the other two velocity components. It can be stated that, in the near-wall region, the agitation of the bubbles is more pronounced in the streamwise and in the spanwise direction and limited in the wall-normal direction. In the core region, instead, the dynamics of the bubbles is almost isotropic in the horizontal (yz) plane and slightly lower in the x-direction. Additionally, the shape of all three profiles matches strongly with the ones of the fluid fluctuation profiles depicted in Fig. 6.6, where also isotropic turbulence was observed in the horizontal plane. The resemblance of fluid and bubble fluctuations, common among all simulations, suggests once more the strong dependency of the bubble motion on the fluid motion and vice versa. When compared to the fluctuation profiles of the *SmMany* case (cf. Fig. 5.15, left) the shapes of the curved are quite different and the turbulence level is much higher for the bubbles in the downward case. In the channel center, the streamwise component is up to roughly three times larger, the wall-normal up to seven times larger and spanwise components and up to roughly four times.

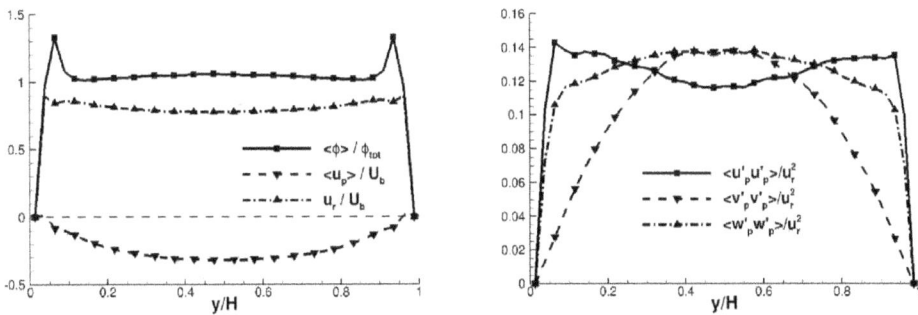

Figure 6.7: Bubble statistics for *SmManyDo*. Left: Void fraction distribution, bubble velocity and relative velocity. Left: Bubble velocity fluctuations in streamwise, wall-normal and spanwise direction scaled with the bubble relative velocity.

The analysis of the mechanisms that influence the mean void fraction distribution $\langle \phi \rangle$, portrayed in Fig. 6.7, left, is now performed and is carried out for $0 < y/H < 0.5$ for clarity. Due to the symmetry of the flow, it also applies for the other half of the channel. The first mechanism discussed here is the shear-induced lift force. For the investigated bubble Reynolds number, it was shown in Sec. 3.2.2 that the lift force acts in the direction of the bubble side where the relative velocity is lower. For the downward configuration the flow velocity is always higher, in magnitude, than the bubble velocity (cf. Fig. 6.5, left, and Fig. 6.7, left). This implies that the relative velocity is lower at the side of the bubble toward the wall, where the fluid velocity is lower. Hence, the lift force moves the bubbles toward the wall. Due to the flatness of the fluid velocity profile (Fig. 6.5), the mean shear is reduced in the core region and increased in a narrow region close to the wall, contributing to the wall peaks of $\langle \phi \rangle$. Bubble collisions, even if slightly higher than in the *SmMany* case, are so limited in number and are not expected to play an important role in the bubble dynamics, as stated in (Michaelides, 2006) for swarms where the total void fraction is very low (as in the present case). The turbophoresis effect was evaluated according to (3.20) and is portrayed in Fig. 6.8. The quantity T_v is always negative and moves the bubbles toward the wall, where the bubble density is higher.

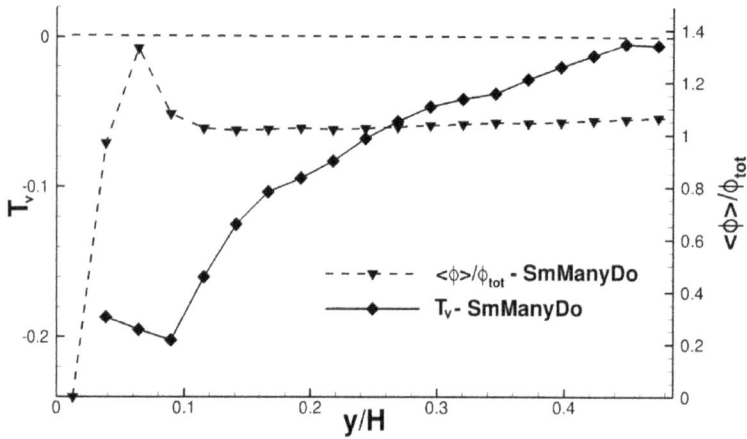

Figure 6.8: Quantification of the turbophoresis effect for simulation *SmManyDo*. Continuous line: wall-normal profile of T_v according to (3.20), evaluated with central difference approximation. Dashed line: void fraction distribution.

6.4   Pair correlation functions

The radial PCF according to (3.22) is shown in Fig. 6.9 and compared with the one of the *SmMany* case and of a random distribution of bubbles. The most probable pair distance is around $3.5r_p$, lower that in the *SmMany* case. Furthermore, the pair frequency for all distances investigated is lower than in the *SmMany* case and larger than the one of RDB. It can be therefore stated that bubble pairs are less probable than in the upward case and that bubbles present a spatial distribution that resembles more the one of RDB. The mechanism of bubble interaction at small scales is, as expected, the same as for bubbles rising in the upward case, related to the lateral low-pressure region around the bubbles. For the investigated void fraction of 2.14%, this mechanism dominates also for bubbles in the downward case, as depicted in Fig. 6.10, where the instantaneous fluid pressure field around a bubble pair is portrayed. The reduced frequency of bubbles pairs compared to the upward case is possibly due to the higher agitation of the bubbles, as previously mentioned. This yields a reduced interaction of the bubbles, especially for the small scales, and the tendency toward a random distribution.

The angular PCF according to (3.23) for the *SmManyDo* case is portrayed in Fig. 6.11. For small distances, the bubbles align mainly horizontal and this feature is due to the lateral pressure zones around rising bubbles, as in the *SmMany* case discussed in Sec. 3.3.5 and portrayed, for the downward configuration, in Fig. 6.10. This tendency is strongly reduced with increasing pair distances and for $6r_p \leq d_{ij} < 9r_p$ the a-PCF is almost indistinguishable from the one of RDB. When compared with the *SmMany* case, the horizontal alignment for small distances is reduced by more than a factor of 2. This suggests that bubbles in the downward configuration follow paths that are more complex and irregular than in the upward case. This implies that the frequency of bubble pairs is reduced and it is an additional proof of the higher agitation experienced by the bubbles.

In conclusion, the simulation of a swarm of bubbles in a downward configuration allows gaining insight into the interaction of bubbles and turbulence for the investigated regime. Liquid turbulence is strongly enhanced with respect both to the unladen flow and the the flow of the same swarm in an upward configuration. The higher turbulence level is caused by the higher agitation of the bubbles, measured with the fluctuation of the bubble velocity components. The higher agitation of the bubbles is also observed when analyzing the mutual position of the bubbles by means of the PCF. Both variants, the radial and the angular one, suggest that the bubble distribution is more similar to the one of random bubbles than the distribution of bubbles in the upward configuration.

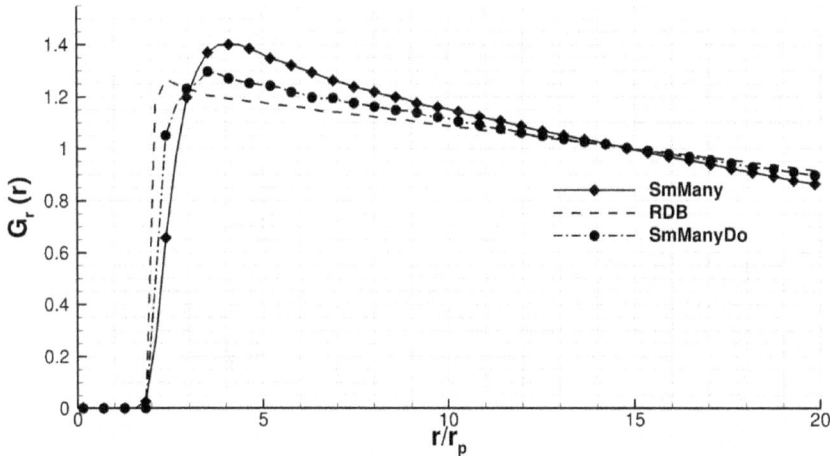

Figure 6.9: Radial PCF according to (3.22) for the *SmManyDo* case, the *SmMany* case and for RDB.

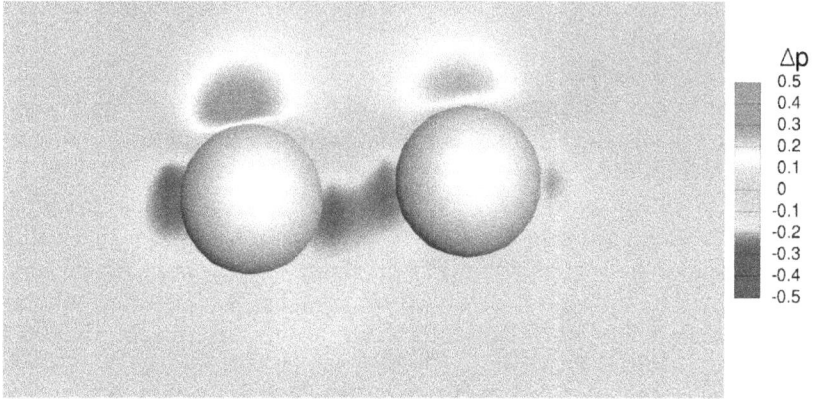

Figure 6.10: Instantaneous pressure field around two rising bubbles, represented on the $(y/H = 0.92)$-plane, where both bubbles centers lie.

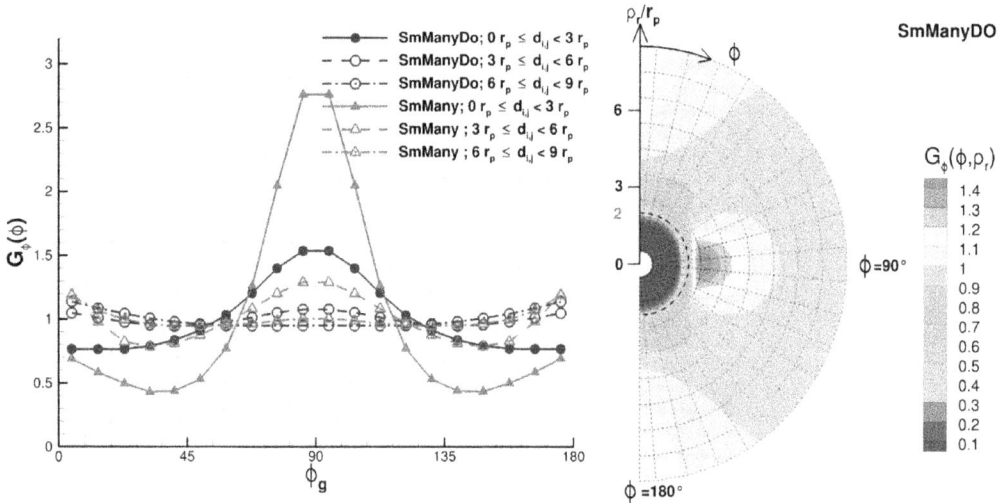

Figure 6.11: Angular PCF according to (3.23) for the *SmManyDo*. Left: Canonical representation and comparison with the *SmMany* case. Right: Two-dimensional representation.

7 Budget of turbulent kinetic energy in bubbly flows

In the previous sections several quantities have been reported for both the carrier and the disperse phase for different flow configurations. It was shown how the bubbles influence the fluid statistics and in particular the fluctuating flow. The turbulent kinetic energy K, in this work referred to as TKE and introduced in Sec. 2.1, is one of the most important quantities in fluid dynamics. It provides information regarding the fluctuating flow field and the turbulence and it is therefore employed in many turbulence models in the framework of RANSE, as described in Sec. 2.1. This observation holds for both single-phase and for two-phase flows flows and in the last decades much effort has been devoted to analyze the mechanisms that influence the fluid field, and hence K, when bubbles are present and. The budget analysis of the transport equation of the TKE of the liquid phase and of each term in the transport equation is the most employed way to gain insight into the complex phenomena involved. Several studies have been so far devoted to this type of analysis and refer to the mathematical formulation of Kataoka (1986) and Kataoka and Serizawa (1989), where the conservation equations of several flow quantities were presented. Fujiwara et al. (2004) experimentally investigated the influence of the bubbles on flow quantities and on the TKE budget in upward pipe flow configurations. Two quantities were evaluated and provided: The production term, related to the gradient of the mean flow, and the dissipation term, responsible for the dissipation of energy at the smallest flow scales. For the regime investigated, these terms are not balanced, independently of the bubble size. Hence, additional mechanisms were supposed to play an important role in the TKE budget represented by an additional term in the budget equation. This term, referred to as the interfacial term, accounts for the energy transfer due to the bubble presence as reported below. Analyzing the flow of bubbles in a similar configuration, Shawkat and Ching (2011) proposed a simplified model for the evaluation of the TKE budget for bubbly flows in pipes and applied such model to their experimental results, neglecting the terms that were not measurable with the experimental apparatus employed. In this model the dissipation of the TKE is evaluated as the sum of a term related to the fluid and another related to the bubbles, whereby the latter was negligible and the one related to the fluid was balanced by the interfacial term. The latter was evaluated by means of a momentum balance between the buoyancy and the drag force in the streamwise direction. Hosokawa et al. (2012) performed experimental measurements to determine the TKE budget for bubbly flows in vertical square ducts. It was found that the production of K was compensated in the whole channel by the dissipation, except in the near-wall region. Due to the available measurement techniques, these authors neglected the pressure-related terms and the interfacial term and assumed an error of around 20% on the TKE budget. The production and dissipation of TKE were also compared to the formulations employed in usual RANSE $K - \epsilon$ models by means of budget equations of both the TKE and the dissipation rate. It was observed that such models are able to capture the general trend of these terms but improvements and additional validation are still needed to develop realistic models. Lelouvetel et al. (2011, 2014) investigated the mechanisms determining the TKE budget for bubbly flows in pipes and strong difference were found between upward and downward flows, employing the simplified model proposed by Tanaka and Eaton (2010). In the upward case the liquid turbulence was reduced by the presence of the bubbles, while it was increased in the downward case. As in (Hosokawa et al., 2012), some terms of the TKE budget were neglected due to available measurement techniques which did not give access to all quantities. In (Lelouvetel et al., 2014), for example, the mean interfacial term of the TKE transport equation was evaluated as the difference between the production and the dissipation term by means of a simplified budget equation. Ilic et al. (2004) and Ilic (2006) employed DNS data of bubbles rising in quiescent fluid[1] to evaluated each term in the TKE budget. Additionally these authors investigated the correctness of the available turbulence models in the framework of bubbly RANSE but the agreement between the evaluation of the terms and the corresponding models was not always satisfactory. The reason proposed by these authors is that such models are developed for highly turbulent flows which is not the case for the configuration investigated in (Ilic, 2006). The work presented by Ilic (2006) is, to the knowledge of the author, the most complete attempt to investigate the TKE budget in bubbly flows, even if the regime investigated by these authors is quite different from the one addressed in this work and hence physical results may only be compared with caution.

This brief overview highlights the need of trustworthy analysis of such phenomena and, for this purpose, the budget analysis is here performed for the simulations discussed in Chaps. 3-6 to investigated how bubbles, under several conditions, modify the turbulence of the carrier phase. At first, the mathematical formulation introduced by Kataoka (1986) and Kataoka and Serizawa (1989) is presented and the conservation equation of the TKE of the liquid phase in bubbly flows is provided. Successively, the numerical procedure is applied to a single-phase flow for validation and to obtain reference data. In Sec. 7.3, the results for the TKE budget analysis for the case *SmMany* are discussed and in Secs. 7.4-7.6 the results for the other simulations are reported. Finally, in Sec. 7.7, the conservation equation of the local instantaneous kinetic energy is introduced and dedicated flow visualizations are provided to investigate the role of each physical mechanism on the turbulence modification induced by the bubbles. A part of the results presented here was provided by Roussel (2014) in the framework of a collaboration between the Technische Universität Dresden, Germany and the Ecole Polytechnique, France.

7.1 Mathematical formulation

A rigorous mathematical formulation of the conservation equations of turbulence quantities in gas-liquid flows was provided by Kataoka (1986) and Kataoka and Serizawa (1989). In this section, the main definitions and equations for the analysis of the TKE of the liquid phase are reported. A pivotal quantity is the liquid phase indicator function Φ defined as

$$\Phi(\mathbf{x}, t) = \begin{cases} 1 & \text{if } (\mathbf{x}, t) \text{ is occupied by liquid} \\ 0 & \text{otherwise.} \end{cases} \tag{7.1}$$

---

[1] "with included effects of column walls" to prevent bubble-wall collision events as stated therein.

Two different averaging operations are required. The first, indicated by a single overline, is the "usual" averaging procedure, i.e. over the whole field. The double overline, instead, indicates the so-called "phase-weighted" average, where only points filled by the liquid are considered. For a given fluid quantity A the definition of the phase-weighted average reads

$$\overline{\overline{A}} = \frac{\overline{A\Phi}}{\overline{\Phi}} \tag{7.2}$$

which requires the average of the liquid indicator function. In the present study, both averaging procedures are performed in the x- and in the z-direction and in time, when not stated otherwise. Fluctuations of fluid quantities are defined with respect to the phase-weighted average as

$$A' = A - \overline{\overline{A}} \tag{7.3}$$

and a similar definition applies for interfacial quantities and corresponding fluctuations, defined as the fluid quantities at the liquid side of the phase boundary, indicated by the subscript L:

$$A'_L = A - \overline{\overline{A_L}} \ . \tag{7.4}$$

According to such formalism, the TKE of the liquid phase is defined as

$$K = \frac{1}{2} \overline{\overline{u'_i u'_i}} \tag{7.5}$$

where, due to the incompressibility of both phase, the density has been neglected (Kataoka and Serizawa, 1989). The fluctuating viscous stress tensor is

$$\tau'_{ij} = \nu \left(\frac{\partial u'_i}{\partial x_j} + \frac{\partial u'_j}{\partial x_i} \right) \ . \tag{7.6}$$

The transport equation of the liquid TKE in bubbly flows according to Kataoka and Serizawa (1989) reads

$$\frac{\partial}{\partial t}(\overline{\Phi} K) + \overline{\overline{u_i}} \frac{\partial}{\partial x_i}(\overline{\Phi} K) = -\frac{1}{\rho} \frac{\partial}{\partial x_i} \left(\overline{\Phi} \, \overline{\overline{p'u'_i}} \right) - \frac{\partial}{\partial x_j} \left(\overline{\Phi} \, \overline{\overline{u'_i u'_i u'_j}} \right) + \frac{1}{\rho} \frac{\partial}{\partial x_j} \left(\overline{\Phi} \, \overline{\overline{u'_i \tau'_{ij}}} \right)$$
$$- \frac{1}{\rho} \overline{\Phi} \, \overline{\overline{\tau'_{ij} \frac{\partial u'_i}{\partial x_j}}} - \overline{\Phi} \, \overline{\overline{u'_i u'_j}} \frac{\partial \overline{\overline{u_i}}}{\partial x_j} - \frac{1}{\rho} \overline{p'_L u'_{L,i} n_i S} + \frac{1}{\rho} \overline{\tau'_{L,ij} u'_{L,i} n_j S} \ . \tag{7.7}$$

The first three terms of the RHS of (7.7) can be collected together and are often refereed to as the transport term, consisting of the transport due to pressure correlation (C_1), due to triple correlation (C_2) and molecular diffusion (C_3), i.e.

$$C = C_1 + C_2 + C_3 = -\frac{1}{\rho} \frac{\partial}{\partial x_i} \left(\overline{\Phi} \, \overline{\overline{p'u'_i}} \right) - \frac{\partial}{\partial x_j} \left(\overline{\Phi} \, \overline{\overline{u'_i u'_i u'_j}} \right) + \frac{1}{\rho} \frac{\partial}{\partial x_j} \left(\overline{\Phi} \, \overline{\overline{u'_i \tau'_{ij}}} \right) \ . \tag{7.8}$$

The fourth term is the viscous dissipation and it represents the rate at which the TKE is dissipated at the small scales

$$\epsilon = -\frac{1}{\rho} \overline{\Phi} \, \overline{\overline{\tau'_{ij} \frac{\partial u'_i}{\partial x_j}}} \ . \tag{7.9}$$

The fifth term is the production term, which accounts for the energy transfer between mean flow and fluctuating flow due to the velocity shear

$$\Pi = -\overline{\Phi} \; \overline{\overline{u_i' u_j'}} \; \frac{\partial \overline{\overline{u_i}}}{\partial x_j} \; . \tag{7.10}$$

The three terms C, ϵ and Π are often referred to as "single-phase" terms, since they are the analogous terms of the transport, dissipation and production term, respectively, in the transport equation of the TKE in single-phase flow, as reported in (Pope, 2000). The sixth and the seventh term of the RHS of (7.7) constitute the interfacial term I

$$I = I_p + I_\tau = -\frac{1}{\rho} \; \overline{p_L' u_{L,i}' n_i S} + \frac{1}{\rho} \; \overline{\tau_{L,ij}' u_{L,i}' n_j S} \; , \tag{7.11}$$

where p_L' is the fluctuation of the fluid pressure, $u_{L,i}'$ the fluctuation of the fluid velocity and $\tau_{L,ij}'$ the fluctuating part of the liquid stress tensor at the liquid side of the phase boundary, respectively. These three quantities have been evaluated according to (7.4). The quantity n_i is the normal vector at the phase boundary directed toward the gas phase (i.e. directed toward the bubble center) and S is the interfacial area concentration (i.e. the interfacial area per unit volume). As defined in (7.11), the interfacial term I accounts for the energy transfer due to the bubble presence and is made of a term related to the pressure fluctuations at the interphase and a term related to the fluctuation of the viscous forces at the interface. According to the definitions introduced above, (7.7) can be rewritten as

$$\overline{\Phi} \frac{DK}{Dt} = \Pi + \epsilon + C + I \; . \tag{7.12}$$

When stationary steady state is reached and average in time is performed over an appropriate period, the material derivative of K is zero, i.e. $DK/DT = 0$, and (7.12) becomes

$$\Pi + \epsilon + C + I = 0 \; . \tag{7.13}$$

Equation 7.13 is the so-called budget equation of TKE for bubbly flows and an evaluation of each term allows gaining insight into the different physical mechanisms that contribute to the TKE.

7.2   Budget of turbulent kinetic energy for single-phase flow

Before addressing the TKE budget in bubbly flows, the budget analysis is performed for the single-phase channel flow introduced in Sec. 3.3. This investigation allows validating the numerical procedure and it provides reference data that will be later referred to when the budget for the bubbly flows is discussed.

The transport equation of the TKE for a single-phase flow can be derived from (7.13) under the condition that $\Phi(\mathbf{x}, t) = 1$ and $S(\mathbf{x}, t) = 0 \; \forall \mathbf{x}, \forall t$ such that (7.13) reduces to

$$\Pi + \epsilon + C = 0 \tag{7.14}$$

as reported, for example, in (Pope, 2000). The TKE budget of the single-phase channel flow was evaluated and the results were compared with the ones of Hoyas and Jimenez

(2008), as portrayed in Fig. 7.1. Velocity components, defined on the staggered grid, were linearly interpolated at the centers of each Eulerian volume and all derivatives involved were approximated by three-point centered finite differences (when not otherwise stated). Due to small difference in the shear Reynolds number ($Re_\tau \approx 169$ for the *Unladen* case and $Re_\tau = 180$ for the reference data) viscous units were employed to scale the results and each term of (7.14) was multiplied by ν/u_τ^4, as in (Hoyas and Jimenez, 2008). A very good agreement between the present results and the reference data is observed so that the numerical procedure for the evaluation of the single-phase terms in (7.13) is validated. For sake of completeness and to allow a direct comparison with the budget analysis of the bubbly flow below, the budget for the unladen channel is also reported here when the terms are scaled with bulk units, i.e. with U_b^3/H, as portrayed in Fig. 7.2.

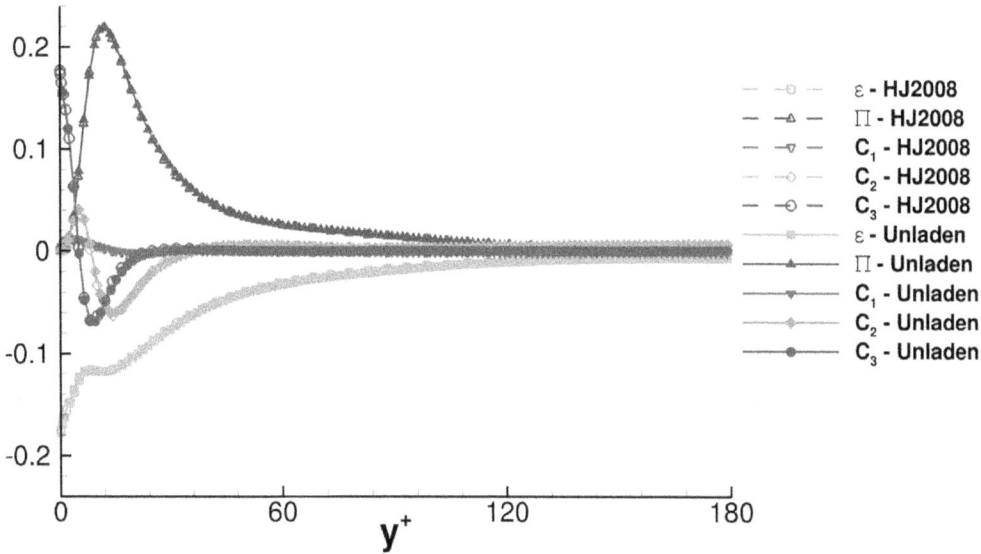

Figure 7.1: TKE budget for simulation *Unladen* and comparison with the results of Hoyas and Jimenez (2008). For the comparison, normalization with wall viscous units was employed: Each term of (7.14) was multiplied by ν/u_τ^4. Results for *Unladen* are averaged on both sides of the channel.

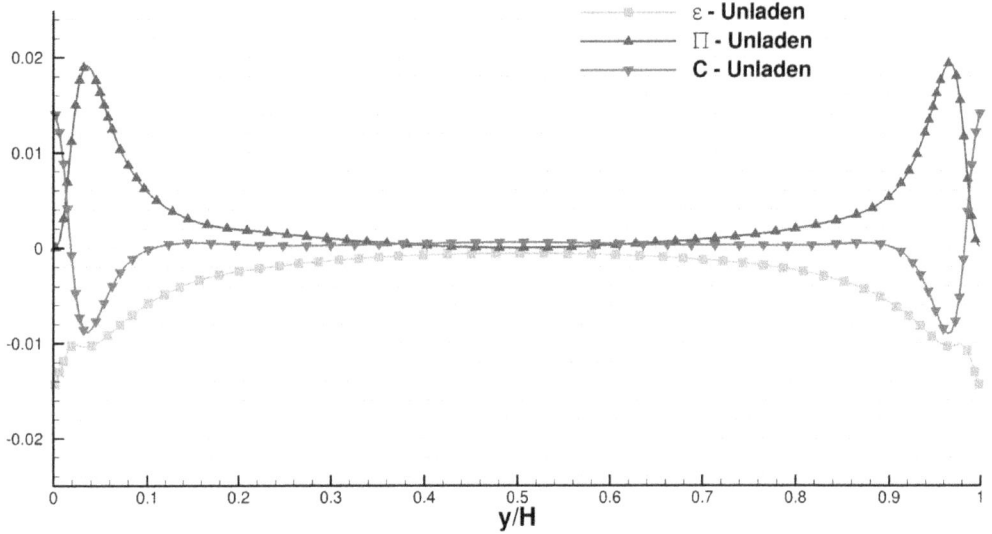

Figure 7.2: TKE budget for simulation *Unladen* in bulk units: Each term is scaled with U_b^3/H.

7.3   Budget analysis for case *SmMany*

Technical approach

When the budget analysis if performed for bubbly flows, particular care has to be devoted some issues. One of this is the evaluation of the pressure field in the vicinity of the phase boundary, which plays a major role in the evaluation of the interfacial term I of (7.13). As mentioned in Sec. 2.3, the coupling between Eulerian and Lagrangian quantities is accomplished by regularized δ-functions and the interpolation stencil spreads over three Eulerian points. Because of these smoothed δ-functions, fluid properties such as pressure and velocity on the Eulerian points in the proximity of the phase boundary may provide inaccurate values when the fluid forces are evaluated, as described in Schwarz (2014). To overcome this problem, a procedure was developed to modify the pressure field on the Eulerian points in the vicinity of the phase boundary. This approach is described in Appendix A and is employed for the evaluation of the pressure field in the analysis of the TKE budget for bubbly flows, both for the term I and for the term C_1. Apart from its dependence from the pressure, the evaluation of the interfacial term I is the most challenging task in the framework of the TKE budget and it is repeated here for convenience:

$$I = I_p + I_\tau = -\frac{1}{\rho}\,\overline{p'_L u'_{L,i} n_i S} + \frac{1}{\rho}\,\overline{\tau'_{L,ij} u'_{L,i} n_j S}\ . \tag{7.15}$$

All quantities involved were defined in Sec. 7.1: Here, we focus on the evaluation of such terms. The quantity S, defined as the interfacial area concentration, is related to the liquid phase indicator function Φ by the following relation (Kataoka, 1986; Kataoka et al., 2012):

$$\frac{\partial \Phi}{\partial x_i} = -S n_i \tag{7.16}$$

and according to Lange (2012) it can be evaluated as

$$S = \sqrt{\left(\frac{\partial \Phi}{\partial x}\right)^2 + \left(\frac{\partial \Phi}{\partial y}\right)^2 + \left(\frac{\partial \Phi}{\partial z}\right)^2}. \qquad (7.17)$$

A preliminary analysis of S has shown that the definition of Φ needs be modified to perform a correct numerical evaluation of S. Hence, the liquid phase indicator Φ is replaced by its numerical counterpart f, i.e. a liquid volumetric fraction. This quantity accounts for the portion of the Eulerian volume occupied by the liquid and can assume each value between 0 (if the discrete volume is entirely occupied by the the gas phase) and 1 (if the discrete volume is entirely occupied by the liquid phase). The evaluation of f is based on the definition of a signed level-set function as proposed by Kempe and Fröhlich (2012a).

Figure 7.3 portrays the interfacial term evaluated according to (7.15) and, since the interfacial term is related to the presence of bubbles, it is not surprising that its shapes resemble strongly the shape of the average bubble volumetric fraction, represented by the dashed line, even if the peaks of I are smoother than to the ones of $1 - \overline{f}$. The interfacial term in Fig. 7.3 and all other terms of the budget equation are scaled with bulk quantities, i.e. with U_b^3/H, and this is done for all pictures reported in this section and in the following[2].

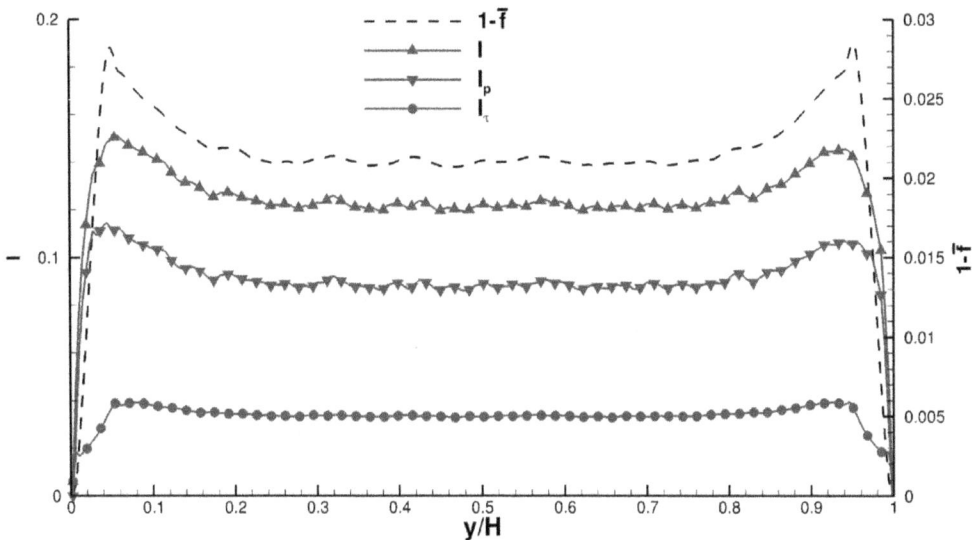

Figure 7.3: Interfacial term for the simulation *SmMany* according to (7.15): Total, pressure-related and viscous stress-related term. The dashed black line represents the average gas volumetric fraction.

It is interesting to note that the subterm I_p related to the pressure is larger by a factor of around three compared to the subterm I_τ, related to the shear stresses. In the channel center, the ratio is 2.67. This means that, for the investigated parameter range, pressure forces provide a stronger contribution to the interfacial term than the forces associated to

---

[2]The only exception is Fig. 7.1, as already reported, where viscous units have been employed to provide a direct comparison with the reference data of Hoyas and Jimenez (2008).

the viscous stress. This is not surprising if one considers the pressure and the friction contribution on the total drag force on a fixed sphere. For $Re_{p,c} = 300$, Johnson and Pater (1999) observed that the ratio between pressure and viscous contribution is around 1.5. Since the drag force is supposed to play the major role in the interfacial term I, the fact that I_p is larger than I_τ matches, at least qualitatively, with the observations in the literature.

A brief consideration is now provided regarding the quantity $1 - \overline{f}$ that represents the gas volumetric fraction. It is clearly related to the void fraction distribution $\langle \phi \rangle / \phi_{tot}$, defined in (3.13) in Sec. 3.3.3. Nevertheless, the approaches yielding these two quantities are somehow different. For the evaluation of $\langle \phi \rangle / \phi_{tot}$, bubbles are considered as whole objects that contributed to $\langle \phi \rangle / \phi_{tot}$ if their centers are inside the wall-normal bin. For $1 - \overline{f}$, instead, the portion of bubble volume associated to each bin is considered and this local approach yields, for example, the smoother peaks of $1 - \overline{f}$ (cfr. Fig. 7.3 and Fig. 3.19 in Sec. 3.3.3). In the framework of the analysis of the TKE budget, the quantity $1 - \overline{f}$ is preferred since it allows a more accurate evaluation of the gas distribution and, hence, of the liquid distribution.

From a technical point of view, some caution must be devoted to the evaluations of I for bubbles approaching the walls or approaching other bubbles since the employment of velocity and pressure values taken from a region outside the physical domain or inside another bubble could lead to numerical inaccuracy. Hence, the interfacial term, i.e. the pressure and the viscous stress on the phase boundary, are evaluated only if at least two discrete Eulerian volumes filled with liquid ($f = 1$) lie between the bubble and the wall or between two bubbles. If this criterion is not fulfilled, pressure and stress fluctuations are locally set equal to zero. This choice may introduce some inaccuracy but, at least, it allows reducing the influence of the points outside the domain and inside the bubbles.

Results

After the technical approach for the evaluation of each term of Eq. (7.12) was discussed in Secs. 7.1 and 7.3, the budget of the TKE accounting for all terms can be found. For the *SmMany* case the budget is portrayed in Fig. 7.4 and now discussed.

The production term Π rapidly increases with the wall distance up to $y_w/H = 0.015$ where it presents its maximum. Then it decreases smoothly for increasing y_w and for $y_w/H = 0.1$ its contribution to the budget is quite limited, around one tenth of the interfacial term I. The shape of Π strongly resembles the one in the unladen flow (cfr. Fig. 7.2) and this is not surprising if one considers that the production term, for the channel flow configuration investigated here, is

$$\Pi \approx -\overline{\Phi} \, \overline{\overline{u'v'}} \frac{\partial \overline{\overline{u}}}{\partial y} \tag{7.18}$$

and both the turbulence stress $\overline{\overline{u'v'}}$ and the fluid velocity $\overline{\overline{u}}$ have similar shapes in the *Unladen* and in the *SmMany* case.

The dissipation term ϵ is highest at the wall (as in the unladen channel) and than rapidly reduces its contribution for increasing y_w. For $y_w/H \approx 0.07$ it presents a local minimum and then, for $y_w/H > 0.25$ the value is fairly constant. With respect to the *Unladen* case, the dissipation is not zero in the core region and this is due to the presence of the bubbles. Indeed, the energy is dissipated the proximity of the bubbles and most of all in the wake region, as reported by Mercado et al. (2012), Mendez-Diaz et al. (2012) and Lelouvetel et al. (2014), to mention but a few, even if the mechanism connecting the dissipation and the wake structures is not fully understood yet. The transport term C, as in the unladen flow,

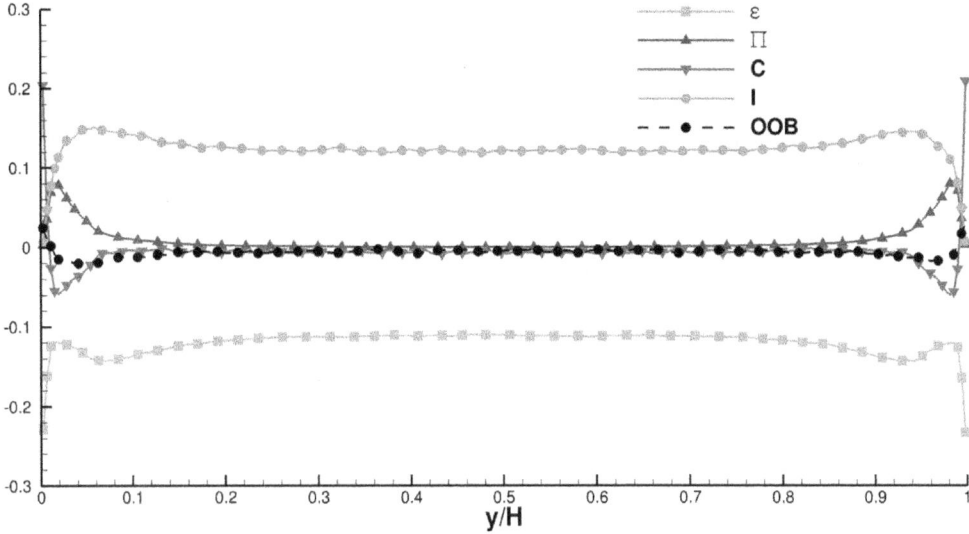

Figure 7.4: TKE budget for simulation *SmMany*.

presents its maximal value at the wall and than rapidly decreases for increasing y_w up to negative values; In this region C is dominated by the shear term C_3 of (7.8) (not shown here). The position of its local minimum is $y_w/H \approx 0.015$, which is the same position of the maximum of Π and of the local minimum of ϵ. For $y_w/H > 0.1$ the transport term is fairly constant with a small negative value. Nevertheless, the shape resembles the one of the transport term in the unladen channel. The interfacial term and its dependence from the bubble distribution for the *SmMany* case have already been reported and discussed in Sec. 7.3. Finally, three regions along the channel width can be observed:

- A region very close to the wall for $y_w/H \lesssim 0.01$ where the positive contributions of Π, of C and of I are balanced by the viscous dissipation.

- An adjacent region for $0.01 \lesssim y_w/H \lesssim 0.25$ where Π and C smoothly decrease with increasing y_w and where I and ϵ first reach maximal and minimal values, respectively, for $y_w/H \approx 0.06$ and than saturate.

- The core regions, for $0.25 \lesssim y_w < 0.5$ where the contributions of Π and C are negligible and the interfacial term is completely balanced by the viscous dissipation.

For the sake of completeness the out-of-balance (*O.O.B.*), defined according to

$$\Pi + \epsilon + C + I + O.O.B = 0 \qquad (7.19)$$

is now discussed. It collects the numerical inaccuracies that may arise in the evaluation of the four terms. As portrayed in Fig. 7.4, this term is almost zero for $0.1 < y/H < 0.9$ and this implies a correct evaluation of the budget equation. Nevertheless, it exhibits small positive values very close to the wall, around 10% of the transport term C (the largest at the wall). This is due to the choice of neglecting the pressure and the viscous term contribution for bubbles approaching the walls and does not influence the overall accuracy of the presented results.

7.4   Influence of the void fraction

The budget analysis for the simulation *SmFew* is now presented and the results for the dilute swarm are compared with the ones of the denser swarm *LaMany*. As discussed in Sec. 4.1 above, the turbulence level of the fluid increases with the void fraction, as depicted in Fig. 7.5, where the profiles of K for all simulations performed are collected. In the channel center, the value of K for *SmFew* is around one fifth of the value for *SmMany*, while the value is around three time the value of K for *Unladen*.

The four terms in the budget equation are portrayed in Fig. 7.6. The shape of the production term is comparable with the one of the *SmMany* case, but, for the dilute swarm, the maxima are decreased by factor 2.6. Analogous considerations can be drawn for the transport term C: The profiles are similar and the maxima are decreases (by factor 4 at the wall and by factor 5.6 for the local minimum). In the dilute swarm, the maxima are further off from the wall and this shift is related to the mean bubble distribution, as already discussed in Sec. 3.3.3 and Sec. 4.3 where the influence of the bubbles on the fluid velocity fluctuations was address. The dissipation term is decreased in the *SmFew* case by a factor of around 7, at the walls and in the core region. It is interesting to note that this factor is very close to the ratio of the total void fraction investigated here, i.e. $\phi_{tot,Many}/\phi_{tot,Few}$=7.5. The dissipation term is strongly related to the bubble presence, except at the wall where it presents a non-zero value also in the unladen case. The shape of the interfacial term is very similar to the shape of the void fraction distribution, as expected. In the core region, where both simulations present an almost constant void fraction distribution, the enhancement of I due to the larger void fraction is around 6, slightly below the ratio of the void fractions.

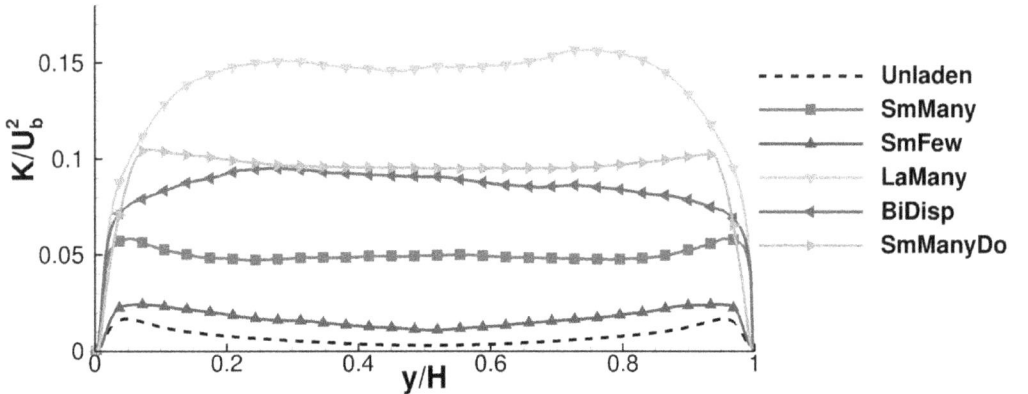

Figure 7.5: Turbulent kinetic energy K according to (7.5) for all simulations performed.

Simulation	*Unladen*	*SmMany*	*SmFew*	*LaMany*	*BiDisp*	*SmManyDo*
K at $y/H = 0.5$	0.0029	0.0494	0.0109	0.1476	0.0906	0.0951
max$\{K\}$	0.0168	0.0584	0.0242	0.1566	0.0950	0.1048

Table 7.1: TKE-values of all performed simulation: Value at the channel center and maximal value.

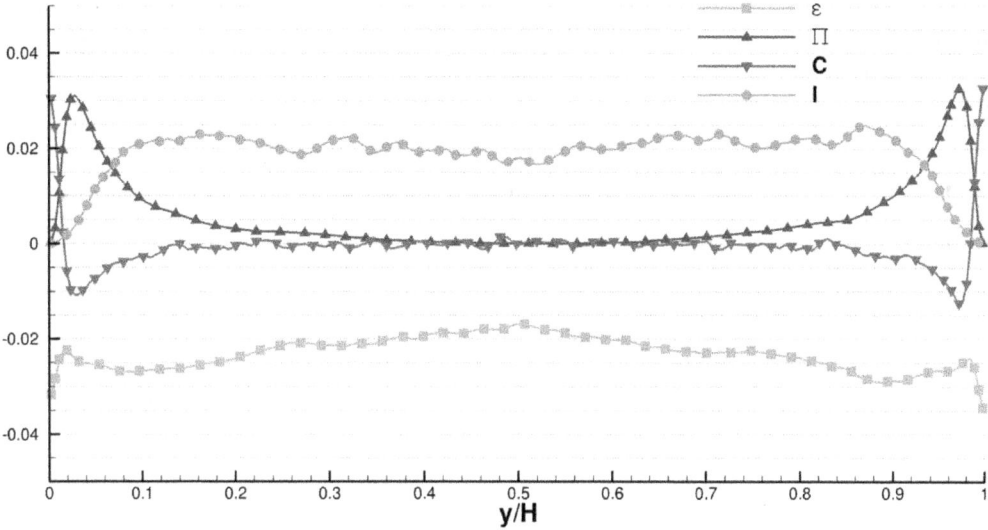

Figure 7.6: TKE budget for simulation *SmFew*.

7.5   Influence of the bubble size

The results of the budget analysis for the swarm of large bubbles, case *LaMany*, are portrayed in Fig. 7.7 and are now discussed. The shape of the production term is quite different with respect to both the *Unladen* and the *SmMany* case and this is due to the modification of the mean fluid velocity profile induced by the bubbles (see Fig. 5.11 in Sec. 5.2.2, left). This implies also that the maximal value at the wall is somewhat lower than in the *SmMany* case, since the term $(\partial\bar{\bar{u}}/\partial y)$ smaller in the near-wall region. The transport term C does not have a minimum close to the wall and, instead, presents a narrow plateau at $y/H \approx 0.02$ and than exhibits small negative values. The slightly irregular attitude of C shows that longer averaging might be beneficial but does not affect the physical meaning here. The same consideration holds also for the interfacial term discussed below. The dissipation term is strongly related to the mean distribution of bubbles: Apart from the wall-normal maxima usual in channel flows, its shape strongly resembles the one of the gas fraction $1 - \overline{f}$. The value of ϵ in the core region is almost twice as high when compared with the *SmMany* case. As for the dissipation term, also the shape of the interfacial term I resembles the one of $1-\overline{f}$ and its maximal value is somewhat lower that the one of the dissipation. For the present case, hence, in the core region the TKE is increased due to the interfacial term and, to a lower extent, to the production term and decreased due to the dissipation term and, to a lower extent, to the transport term.

When the bidisperse swarm is considered, similar observations can be made. As displayed in Fig. 7.8, the profiles of the four terms lie between the corresponding ones of the *SmMany* and of the *LaMany* cases. The shape of the interfacial terms shows a stronger similarity with the average distribution of the large bubbles (see Fig. 5.23 in Sec. 5.3) than of the small ones. It can hence be deduced that, for the investigate regime, the large bubbles provide a larger contribution to the interfacial term than the small ones, even if the total void fraction is evenly distributed. Although the total surface of small bubbles is larger than the total surface of the larger ones, the relative velocity is different, an this is what is decisive. This feature is not so pronounced for the dissipation term, although a flat (local) minimum is present in the channel center.

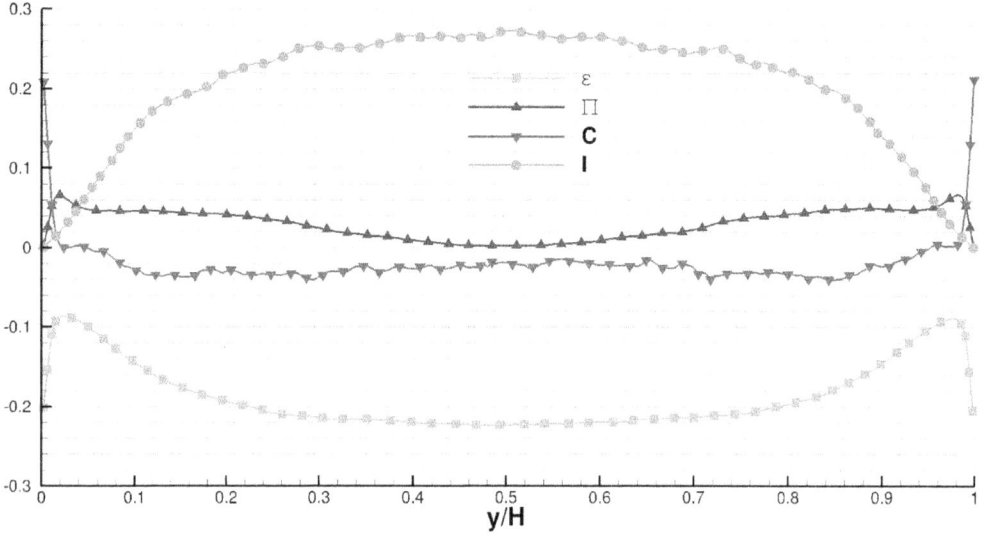

Figure 7.7: TKE budget for simulation *LaMany*.

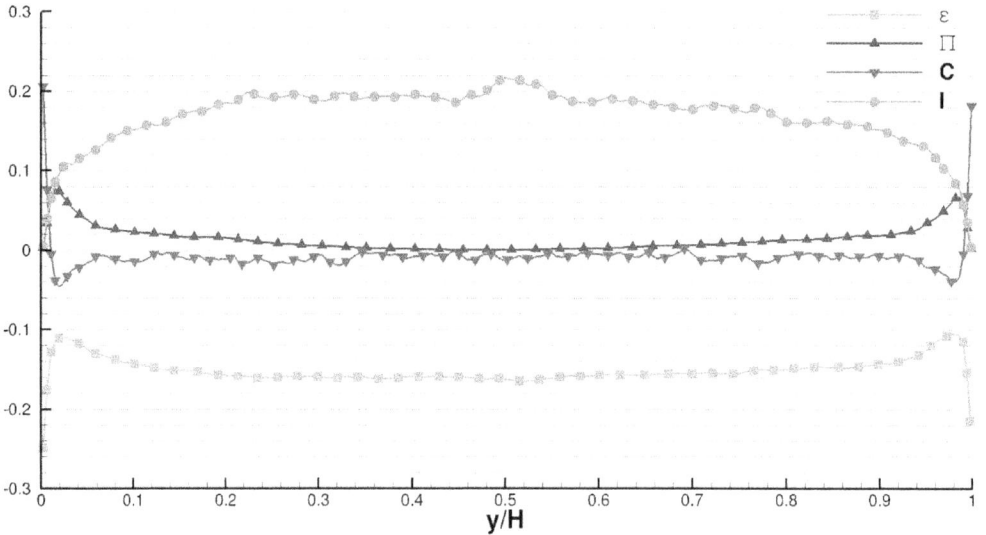

Figure 7.8: TKE budget for simulation *BiDisp*.

7.6   Influence of the flow direction

The picture is quite different if one compares the *SmMany* and the *SmManyDo* case to address the influence of the flow direction on the TKE budget. As already discussed in Secs. 6.1, in the downward case the turbulence enhancement due to the bubble presence is larger than the one induced in the upward case and this is shown in Fig. 7.5 where the profiles of the TKE are portrayed. The TKE is increased by almost a factor of 2 in the core region, where both simulations present a fairly constant bubble distribution. For the *SmManyDo* case, both the maxima of the void fraction distribution and of the TKE are further away from the wall. The budget analysis for the *SmManyDo* in Fig. 7.9 shows that the relation between the four terms is independent of the flow direction: The interfacial term and, to a lower extent, the production term provide a positive contribution to K, while for the dissipation term and, to a lower extent, the transport term the contribution to the budget is negative. Nevertheless, some differences may be noticed. For the production term, the maxima are higher for the *SmMany* case and the overall lower magnitude for the *SmManyDo* case is due to the lower turbulent shear stress $\langle u'v' \rangle$, as shown in Fig. 6.6. The bump of Π for the *SmManyDo* case at $y_w/H \approx 0.05 \approx d_p$ is a consequence of the bump in the turbulent shear stress profile. It can hence be stated that the production term accounting for the transfer of energy from the mean flow to the fluctuation flow is smaller in the downward case due to the reduced correlation between streamwise and wall-normal fluctuations, quantified by the turbulent shear stress $\langle u'v' \rangle$, as discussed in Sec. 6.3. Regarding the transport term C, it can be observed that the shape is quite similar but, except at the wall, C is higher (in magnitude) in the downward case. For the dissipation term the shape of the profiles is also quite similar. At the wall, the dissipation is higher for the upward case, although the wall shear-stress is higher in the downward case. In the core region, the magnitude of the dissipation term is higher for the *SmManyDo* case and this is due to the higher agitation induced by the bubbles on the fluid: The energy is produced and immediately dissipated near the bubbles, as reported also in (Lelouvetel et al., 2014). The interfacial term follows the profile of the void fraction in both case and is somewhat higher in the downward case to balance the somewhat higher (in magnitude) dissipation term and the transport term.

7.7   Local instantaneous turbulent kinetic energy

In the present section, the DNS data are employed to obtain local and instantaneous information regarding the influence of the bubbles on the fluctuating fluid velocity. This analysis can be considered as the counterpart of the investigation performed in the previous sections, when the influence of the bubbles on the turbulence of the liquid phase was represented by means of averaged quantities, such as the TKE. To analyze the local and instantaneous impact of bubbles, the local instantaneous kinetic energy is defined as

$$\widetilde{K} = \frac{1}{2}\, u_i' u_i' \,. \qquad (7.20)$$

Note that this quantity implies the definition of averaged quantities, i.e. of averaged velocity components, otherwise no fluctuation could be defined. The conservation equation of $\widetilde{K}$

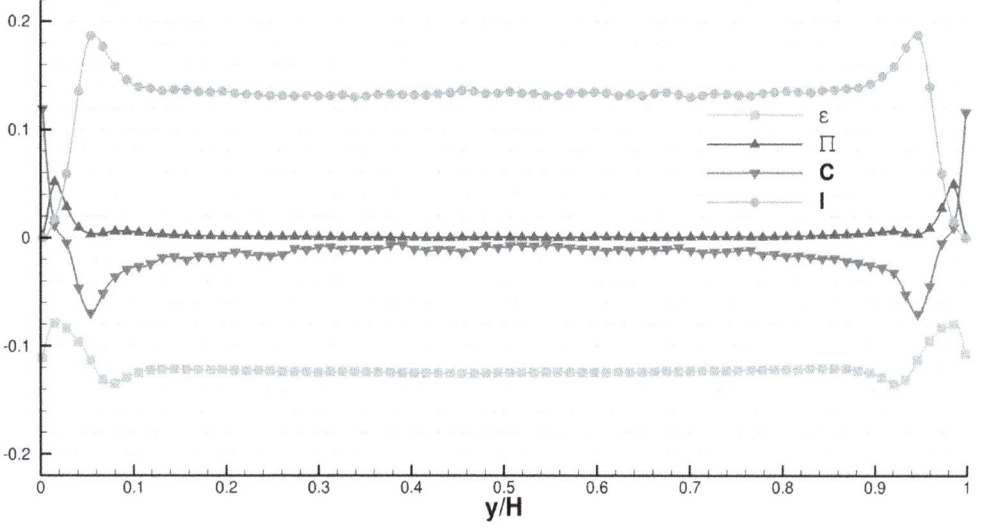

Figure 7.9: TKE budget for simulation *SmManyDo*.

reads

$$\Phi\frac{\partial}{\partial t}\widetilde{K} + \Phi\frac{\partial}{\partial x_j}\left(\widetilde{K}\,\overline{\overline{u_j}}\right) = -\Phi\frac{1}{\rho}\frac{\partial}{\partial x_i}\left(p'u_i'\right) + \Phi\frac{1}{\rho}\frac{\partial}{\partial x_j}\left(\tau_{ij}'u_i' + \overline{\overline{u_i'u_j'}}u_i'\right) +$$

$$-\Phi\frac{1}{\rho}\left(\tau_{ij}' + \overline{\overline{u_i'u_j'}}\right)\frac{\partial u_i'}{\partial x_j} - \Phi\frac{\partial}{\partial x_j}\left(\overline{\overline{u_i'u_j'}}u_j'\right) - \Phi u_i'u_j'\frac{\partial\overline{\overline{u_i}}}{\partial x_j} + \quad (7.21)$$

$$+\frac{\Phi}{\overline{\Phi}}\frac{1}{\rho}\overline{u_i'\left(p'n_i - \tau_{ij}'n_j - \overline{\overline{u_i'u_j'}}n_i\right)S} - \frac{\Phi}{\overline{\Phi}}\frac{1}{\rho}p'\,\overline{u_i'n_iS}\,,$$

as formulated by Kataoka and Serizawa (1989). The reader may notice that, when applying the phase-average operator defined in (7.2), some terms are equal to zero, e.g.

$$\overline{\overline{\Phi\frac{1}{\rho}\frac{\partial}{\partial x_j}\left(\overline{\overline{u_i'u_j'}}u_i'\right)}} = \overline{\Phi}\frac{1}{\rho}\frac{\partial}{\partial x_j}\left(\overline{\overline{u_i'u_j'}}\,\overline{\overline{u_i'}}\right) = 0 \quad \text{since} \quad \overline{\overline{u_i'}} = 0 \quad\quad (7.22)$$

and eventually (7.7) is found. Here, instead, the focus is on the terms of (7.21) to provide information on the local instantaneous modification of $\widetilde{K}$ induced the bubbles. A grouping of the terms of (7.21) is introduced in a similar manner as for (7.7). The local instantaneous production $\widetilde{\Pi}$ is defined as

$$\widetilde{\Pi}(x,y,k,t) = -\Phi u_i'u_j'\frac{\partial\overline{\overline{u_i}}}{\partial x_j} \quad\quad (7.23)$$

and analogously the local instantaneous dissipation $\widetilde{\epsilon}$ is defined as

$$\widetilde{\epsilon}(x,y,k,t) = -\Phi\frac{1}{\rho}\tau_{ij}'\frac{\partial u_i'}{\partial x_j}\,. \qu\quad (7.24)$$

The local instantaneous transport $\widetilde{C}$ is defined as

$$\widetilde{C}(x,y,k,t) = -\Phi\frac{1}{\rho}\frac{\partial}{\partial x_i}\left(p'u_i'\right) + \Phi\frac{1}{\rho}\frac{\partial}{\partial x_j}\left(\tau_{ij}'u_i' + \overline{\overline{u_i'u_j'}}u_i'\right) - \Phi\frac{\partial}{\partial x_j}\left(\overline{\overline{u_i'u_j'}}u_j'\right) - \Phi\frac{1}{\rho}\left(\overline{\overline{u_i'u_j'}}\right)\frac{\partial u_i'}{\partial x_j}$$

$$(7.25)$$

where the last term is collected in the transport term even if slightly different from the others terms. Eventually, all terms related to the phase boundary are grouped in the local instantaneous interfacial term $\widetilde{I}$ defined as

$$\widetilde{I}(x,y,k,t) = -\frac{\Phi}{\overline{\Phi}}\frac{1}{\rho}\overline{p'\,\overline{u_i'n_iS}} + \frac{\Phi}{\overline{\Phi}}\frac{1}{\rho}\overline{u_i'\left(p'n_i - \tau_{ij}'n_j - \overline{\overline{u_i'u_j'}}n_i\right)S}\,. \qquad (7.26)$$

This analysis was performed for all simulations presented here and, as for the budget analysis, results for the reference case *SmMany* are presented at first. Figure 7.10, left, portrays the local production term $\widetilde{\Pi}$: It exhibits the largest values in the vicinity of bubbles rising near the walls where a local interaction between bubbles and near-wall turbulence is observed. This yields both positive and negative values of the instantaneous production term. It means that, in the near-wall region, energy is transferred from the mean field to the fluctuating field and vice versa. Bubbles rising in the center region provide a marginal contribution to the production term since the mean velocity gradient is almost zero here, i.e. $(\partial\overline{\overline{u}}/\partial y) \approx 0$. Figure 7.10, right, portrays the local dissipation term $\widetilde{\epsilon}$ for the *SmMany* case. The dissipation is always negative and exhibits non-zero values only in the vicinity of the bubbles and very close to the walls. The highest values of the dissipation term are observed above and on the side of the bubbles, where the velocity gradients are largest. It can be observed that in the wake region, the shape of the dissipation is strongly related to the shape of the wake itself and that, as in the vicinity of the bubble, the dissipation is large where velocity gradients are higher.

The transport term $\widetilde{C}$, shown in Fig. 7.11, left, is also related to the bubble presence, but exhibits different values in the vicinity of the bubbles. Above the bubbles, the transport term is positive. This means that, in these regions, it is responsible for the energy transfer from the mean flow to the fluctuating part. Below the bubbles, instead, it exhibits negative values and drains energy from the fluctuating field. Although the values of $\widetilde{C}$ are lower than $\widetilde{\epsilon}$, it may be observed that in the core of the wake the energy is transported from the fluctuating field to the mean filed, i.e. fluctuations induced in the bubbles wake are dispersed in the channel. The instantaneous interfacial term $\widetilde{I}$ is portrayed in Fig. 7.11, right, for sake of completeness and it presents non-zero values only in the proximity of the phase boundary where $S(i,j,k,t) \neq 0$, as expected.

As mentioned above, the same analysis was performed also for all other simulations presented in this work but only selected flow visualizations are reported here. Figure 7.12 portrays the production term (left) and the dissipation term (right) for the bidisperse swarm *BiDisp* where the influence of the two bubble classes may be appreciated. The production term is also related to the bubbles, but it exhibits larger values in the core region due to the modification of the mean velocity profile induced by the bubbles (see Fig. 5.11 in Sec. 5.2.2) as already observed in the TKE budget in Fig. 7.8. This feature is even more pronounced in the *LaMany* case (not shown here) where the mean fluid velocity profile differs even more from the turbulent velocity profile of single-phase channel flow. Figure 7.12, right, shows the simultaneous contribution of small and large bubbles on the the local instantaneous dissipation field. As expected, the contribution of the large bubbles extends to larger distances due to the stronger modification of the fluid velocity induced by the large bubbles and by their wakes. Figure 7.13 portrays the same analysis for the downward case *SmManyDo*. No remarkable difference is observed for the production term with respect to the upward case *SmMany* :$\widetilde{\Pi}$ is large where bubbles interact with the walls and zero in the core region. The dissipation term, instead, presents some differences which are mainly related to the motion of the bubbles. In the upward configuration, Fig. 7.10, the influence of the bubbles on the dissipation was observed mainly in the region below the bubbles, due to the almost straight paths of the bubbles. In the downward configuration, instead, the influence of the bubbles is more irregular. This is due to the higher agitation of the bubbles. As depicted in Fig. 7.10, the higher dynamics in the wall-normal direction yields larger dissipation regions that present more irregular shapes, for example in oblique regions below the bubbles.

For sake of completeness, Appendix B reports the same analysis, i.e. governing equation and visualizations, for a single-phase channel flow.

In conclusion, a detailed analysis of the budget equation of the TKE has been provided for the different cases investigated to gain insight into the turbulence modification induced by the bubbles. To the knowledge of the author, this is the first time where such analysis was performed for bubbles in turbulent flows. For all configurations presented, the bubbles act as sources for the TKE and this increase is balanced by the larger dissipation when compared to the unladen case. Additionally, the analysis of each term of the transport equation of the local instantaneous kinetic energy allows to address the local and instantaneous influence of the bubbles on the different mechanisms related to the turbulence modification.

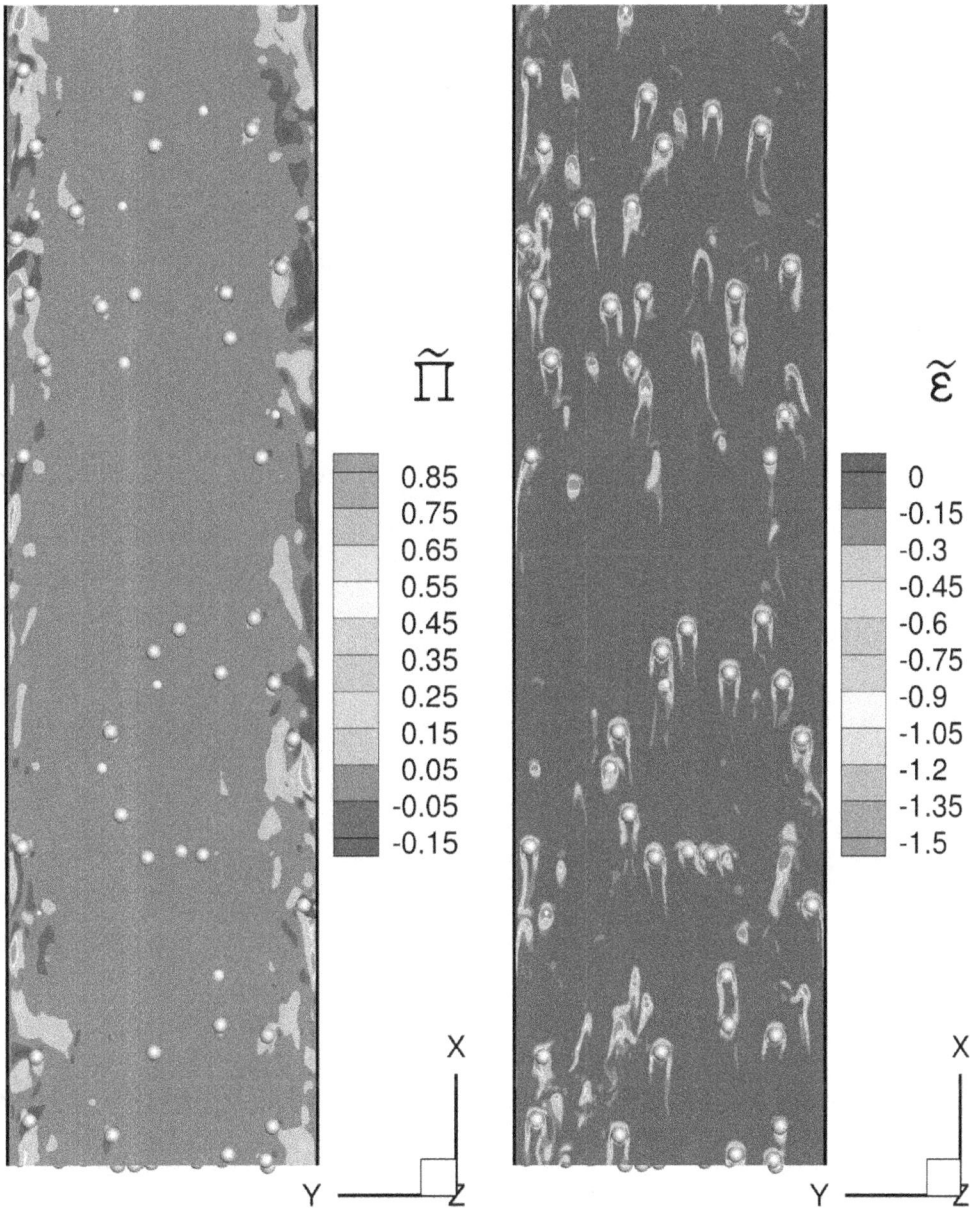

Figure 7.10: Local instantaneous production term $\tilde{\Pi}$ (left) and dissipation term $\tilde{\epsilon}$ (right) according to (7.23) and (7.24), respectively, for case *SmMany*. Contours represented on a vertical wall at $z/H = L_z - r_p$. The whole channel width L_y is portrayed and the walls are represented by the black vertical lines.

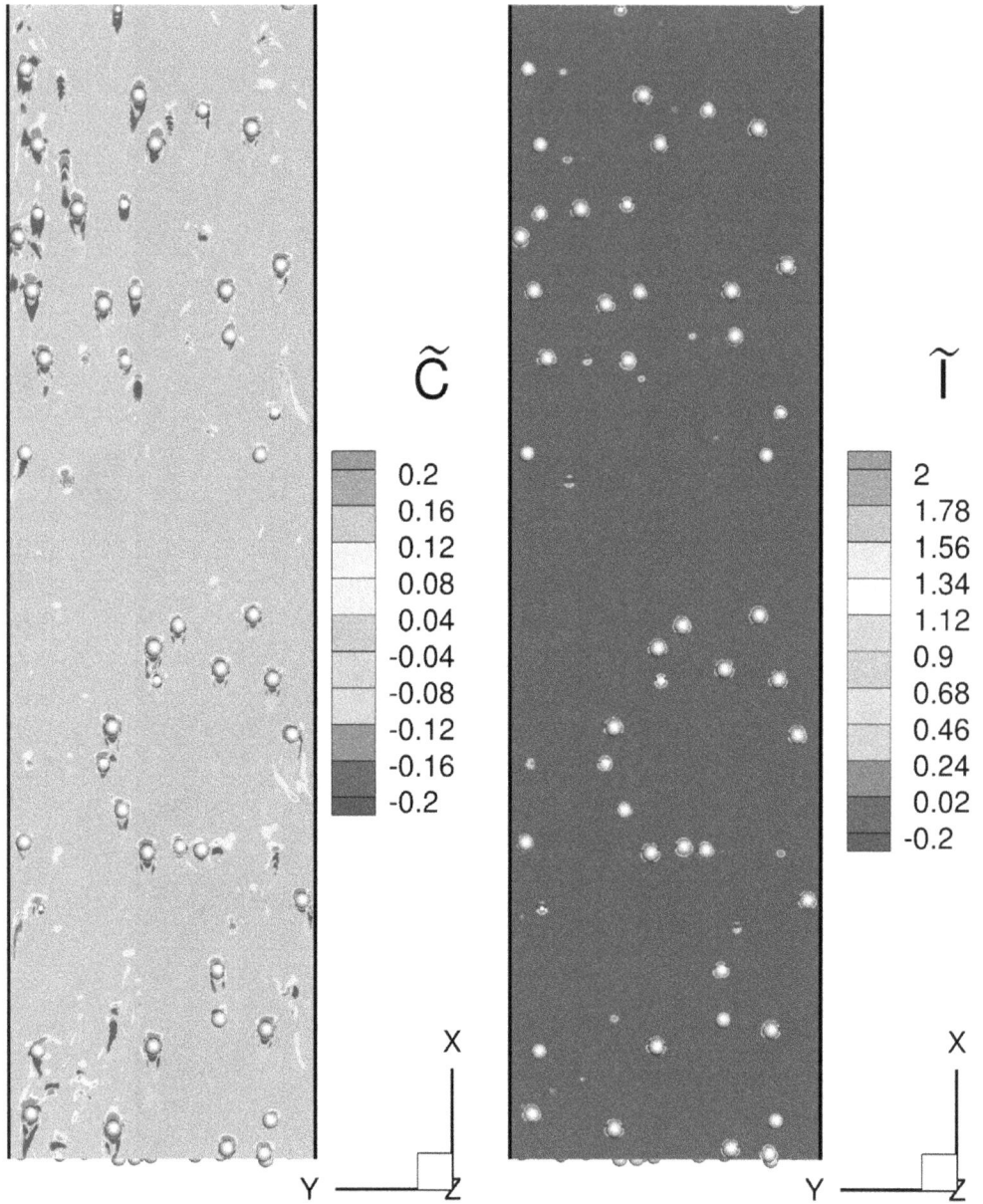

Figure 7.11: Local instantaneous transport term $\tilde{C}$ (left) and interfacial term $\tilde{I}$ (right) according to (7.25) and (7.26), respectively, for case *SmMany*. Contours represented on a vertical wall at $z/H = L_z - r_p$. The whole channel width L_y is portrayed and the walls are represented by the black vertical lines.

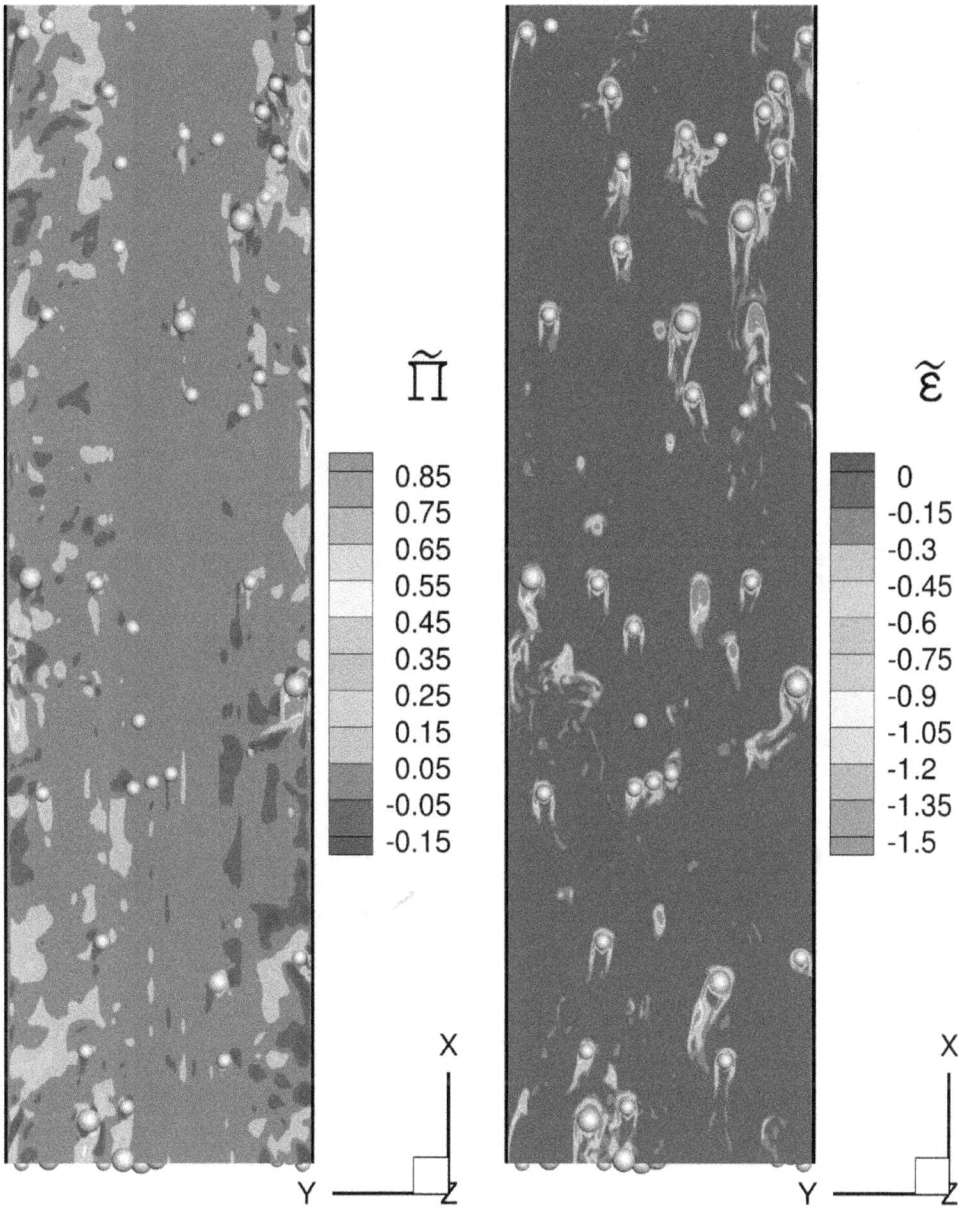

Figure 7.12: Local instantaneous production (left) and dissipation term (right) according to (7.23) and (7.24), respectively, for case *BiDisp*. Contours represented on a vertical wall at $z/H = L_z - r_{p,LA}$. The whole channel width L_y is portrayed and the walls are represented by the black vertical lines.

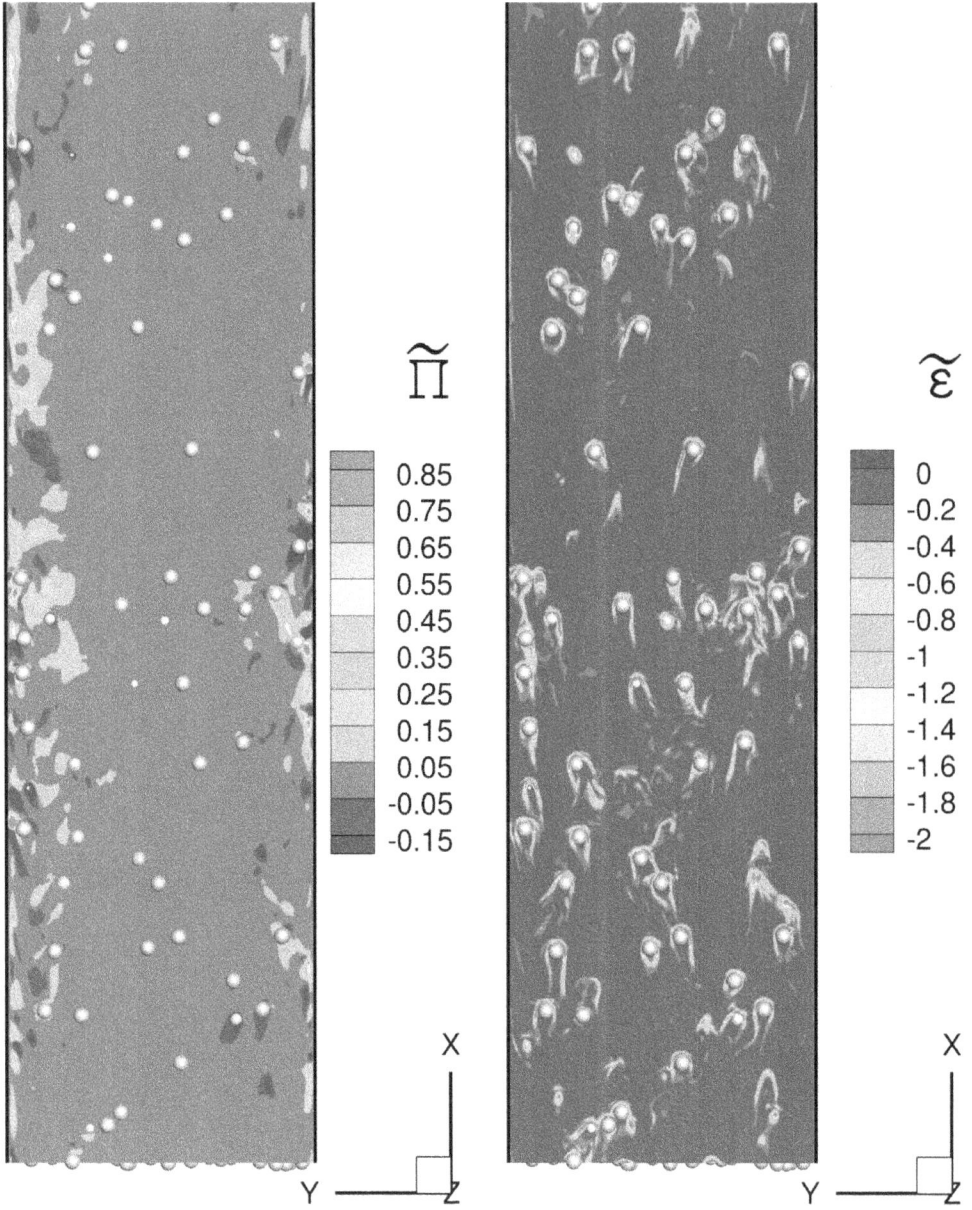

Figure 7.13: Local instantaneous production term $\widetilde{\Pi}$ (left) and dissipation term $\tilde{\epsilon}$ (right) according to (7.23) and (7.24), respectively, for case *SmManyDo*. Contours represented on a vertical wall at $z/H = L_z - r_p$. The whole channel width L_y is portrayed and the walls are represented by the black vertical lines.

8 Heat transfer in bubbly flows

One of the main aspects in the framework of two-phase flows is the heat transfer between the two phases which plays a significant role in many industrial applications as well as in several environmental systems. The involved phenomena strongly depend both on the hydrodynamics and on the thermal properties of the mixture. In the case of air bubbles in water, as investigated in the present work, the large difference of the values of the thermal conductivity of the two phases plays a significant role and numerical models have to be employed to account for it. To address such problems the code PRIME was improved to allow the analysis of the temperature field of the two-phase mixture and such effort is reported here. First, a transport equation for the temperature was implemented and validated for an unladen flow in a channel configuration. Afterwards, the implementation of the models for the correct application of a *thermal* boundary condition at the phase boundary is discussed and such models are validated for the flow around a fixed sphere under several conditions. Finally, results are presented regarding a the simulation where the focus is on the heat transfer between the two phases, i.e. between the fluid and the bubbles and between the mixture and the walls containing the flow.

8.1 Temperature transport equation

8.1.1 Implementation

From the conservation equation of the thermal energy $E = \rho C_p T$ for an incompressible fluid (Baehr and Stephan, 2004), the trasport equation for the temperature can be drived as

$$\frac{\partial T}{\partial t} + \mathbf{u} \cdot \nabla T = a \nabla^2 T \tag{8.1}$$

where the thermal diffusivity is defined as

$$a = \frac{k}{\rho C_p} \tag{8.2}$$

where k is the thermal conductivity and C_p the specific heat capacity. Here, constant physical properties have been assumed. The numerical discretization of (8.1) matches the one of the Navier-Stokes equations (see Chap. 2) and is briefly recalled here. The semi-discrete form of (8.1) reads

$$\frac{T^k - \tilde{T}^k}{\Delta t} + \frac{\tilde{T}^k - T^{k-1}}{\Delta t} = -\gamma_k \, \nabla \cdot (\mathbf{u}T)^{k-1} - \zeta_k \, \nabla \cdot (\mathbf{u}T)^{k-2}$$
$$+ \frac{1}{2}(2\alpha_k) a \nabla^2 T^{k-1} + \frac{1}{2}(2\alpha_k) a \nabla^2 T^k \tag{8.3}$$

where $\tilde{T}^k$ is an intermediate temperature field between the two consecutive Runge-Kutta (RK) steps $k-1$ and k. The advection in time of the convective term of (8.1) is performed by a three-step RK scheme and for each sub-step $k = 1, 2, 3$ the diffusive term of (8.1) is discretized by means of a Crank-Nicholson scheme. The solution procedure can be split in two parts for each of the RK sub-step and the first part is solved explicitly as

$$
\begin{aligned}
\tilde{T}^k = T^{k-1} &-\gamma_k \, \Delta t \, \nabla \cdot (\mathbf{u}T)^{k-1} \\
&-\zeta_k \, \Delta t \, \nabla \cdot (\mathbf{u}T)^{k-2} + (2\alpha_k)\Delta t \, a\nabla^2 T^{k-1}
\end{aligned}
\tag{8.4}
$$

where $\tilde{T}^k$ is the intermediate temperature field for the sub-step k and coefficient $\alpha_k, \gamma_k, \zeta_k$ are the Runge-Kutta weights for time integration, as reported in (Kempe and Fröhlich, 2012a). The second part of (8.1), instead, is solved explicitly as

$$
\left[-\frac{T^k}{(\alpha_k)\, a\Delta t} + \nabla^2 T^k \right] = -\frac{1}{a\,(\alpha_k)} \left(\frac{\tilde{T}^k}{\Delta t} \right) + \nabla^2 T^{k-1} \, .
\tag{8.5}
$$

where $T^{k-1} = T^n$ for the first sub-step $k = 1$ and $T^k = T^{n+1}$ for the last sub-step $k = 3$ while n is the time step. All fluxes can be computed by means of central-difference schemes except at the domain boundaries. Regarding the coupling between velocity and temperature field, for each sub-step (8.4) and (8.5) are solved after a divergence-free velocity field is obtained after the pressure-correction equation. This scheme applies both for single-phase and for two-phase flows.

8.1.2 Validation for single-phase flow

The transport of the temperature as a passive scalar in a turbulent channel flow was simulated and compared with data from the literature to validate the implementation of the transport of heat. The simulation was already introduced in Chap. 3 and labeled *Unladen* and in this section the focus is given to the temperature field.

The Prandtl number ν/a was set to 1, periodic conditions were applied for the temperature in the streamwise and spanwise direction, while Dirichlet BC were applied at the two walls:

$$
T(x, y/H = 0, z) = T_{ref} \quad \text{and} \quad T(x, y/H = 1, z) = -T_{ref} \, .
$$

The instantaneous temperature field is portrayed in Fig. 8.1 by means of the contour plot on the $z/H = L_z/2$-plane. As for the statistics regarding the velocity field, the statistical quantities for the temperature field were collected for around $T_{st} = 400 T_b$ after a stationary state was reached. The mean temperature profile and the mean temperature fluctuations are portrayed in Fig. 8.2: Good agreement is found between the performed simulation and the DNS results of Denev et al. (2008). The wall-normal position of the maximum of $\langle T'T' \rangle$ is slightly different due to the slightly different shear Reynolds number, which is $Re_\tau = 180$ for the simulation in (Denev et al., 2008) and $Re_\tau \approx 169$ for the *Unladen* case.

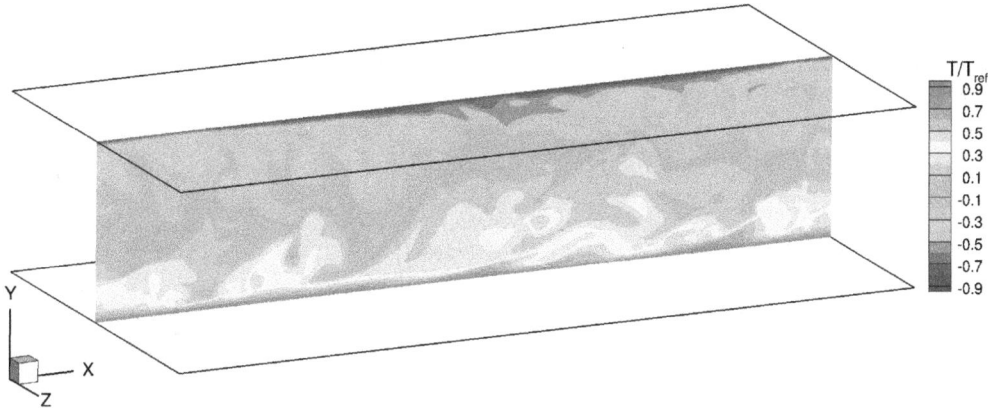

Figure 8.1: Instantaneous temperature field in the unladen channel flow: temperature profile on the $z/H = L_z/2$-plane.

8.2 Numerical methods for thermal problems in two-phase flows

8.2.1 Introduction

Several approaches have been proposed in the literature to impose thermal boundary conditions (BC) on the phase boundary (PB) in the framework of two-phase flows. Some of these approaches are now discussed, and afterwards the choice of the implemented method in the code PRIME is motivated. Zhang et al. (2008) developed an IBM based on a direct forcing approach to simulate the flow around a fixed cylinder with heat transfer from the cylinder surface. The coupling between Eulerian and Lagrangian quantities is based on a bilinear interpolation and both Dirichlet and Neumann BC on the cylinder surfaces are simulated. For the Neumann BC, an additional layer of Lagrangian points is defined in the fluid region at a certain radial distance from the cylinder surface. The temperature on the points in the cylinder surfaces is extrapolated by means of a linear approximation in the radial direction employing the temperature value at the additional points. Wang et al. (2009) presented an IBM for the simulation of heat transfer problems in two-dimensional cases and for fixed objects, and this method is similar to the one employed in the present work for the coupling of the velocity field. The interpolation of the Eulerian quantity, i.e. velocity and temperature, and the spreading of the Lagrangian ones is accomplished by means of the discrete delta functions proposed by Griffith and Peskin (2005). Additional loops are employed to reduce the numerical error between desired and interpolated temperature on the phase boundary in a similar manner as the one proposed in Kempe and Fröhlich (2012a) for the velocity. In (Wang et al., 2009) only Dirichlet BC are investigated. Pan (2010, 2012) proposed a Ghost Cell Method to investigate the heat transfer around fixed objects. This method is based on the definition of the so-called "ghost points" inside the immersed objects, which are employed to impose the desired BC on the phase boundary. The coupling between Lagrangian and Eulerian quantities is accomplished by bi-linear interpolation schemes and the

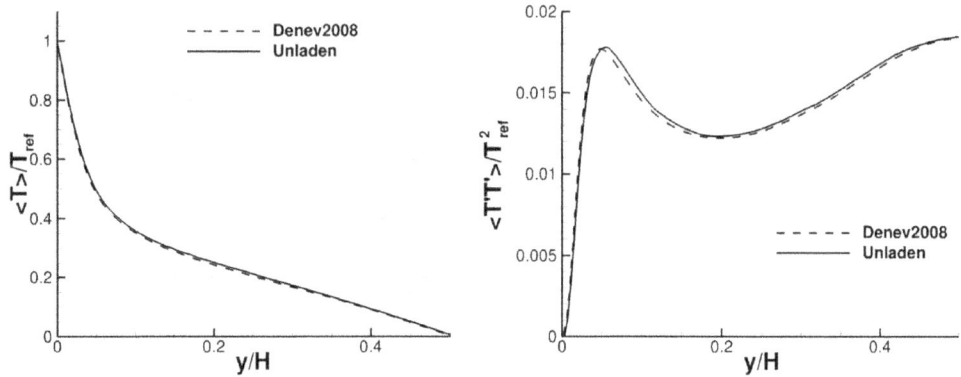

Figure 8.2: Statistics of the temperature field in the unladen channel flow: mean temperature profile (left) and mean temperature fluctuations (right). Comparison with results of Denev et al. (2008).

method is able to impose Dirichlet as well as Neumann BC. In the latter case, an additional layer of Lagrangian points needs to be defined and two variants are proposed. The first is based on a linear approximation of the temperature along the radial direction, employing the temperature values on the additional layer of Lagrangian points, similarly to (Zhang et al., 2008). The second variant is based on a second order approximation, where temperature and temperature gradients on the additional Lagrangian points are employed. Deen and Kuipers (2013) employed a VOF approach to investigate the heat transfer between a hot wall and a cold fluid in the presence of a rising bubble. The numerical methodology employed relies on the work presented by Van Sint Annaland et al. (2005) and is completed by a transport equation of the thermal energy in the form of a transport equation of the temperature. Results regarding the heat transfer at the wall are reported for several cases, such as for the flow of single bubbles as well as of bubble pairs. To investigate the non-isothermal flow around arrays of particles, Tavassoli et al. (2013) employed an IBM based on a direct heating approach and convergence loops similar to Wang et al. (2009) and Kempe and Fröhlich (2012a), where the heating process is repeated until a certain convergence criterion between evaluated and desired temperature is satisfied. For the momentum forcing, a procedure similar to the one proposed by Uhlmann (2005) is employed. This method is applied, among others, for the simulation of the the flow through a static array of particles with heat transfer, obtaining very good agreement with empirical correlations in the literature. To investigate a similar problem, i.e. the heat transfer around a cluster of fixed particles, Xia et al. (2014) employed a GCM based on a third-order approximation of field quantities and these values are evaluated by the governing transport equation where the matrix of the coefficients of the systems of linear equations is modified to account for the desired boundary condition. This approach was proposed by Seo and Mittal (2011) and employed for both Dirichlet and Neumann thermal BC. The numerical accuracy is found to be comparable to the one obtained with an IBM (similar to the one used by Wang et al. (2009) and Tavassoli et al. (2013)) but employing a smaller number of grid points for the discretization of the immersed objects.

This brief overview highlights the several possibilities that can be employed for the simulation of thermal problem in two-phase flows. The GCM proposed by Pan (2010) is expected to provide very good results for fixed objects, while the IBM of Wang et al. (2009) and the VOF of Deen and Kuipers (2013) are expected to strongly reduce the numerical inaccuracy when moving objects are simulated. In relation to the simulation of rising bubbles, the method proposed by Wang et al. (2009) seems to be the best choice, also due to the extremely similar method employed for the coupling of the velocity field used in the present work. To this end, an IBM has been implemented in the code PRIME that is able to handle Dirichlet as well as and Neumann BC. To the knowledge of the author, this is the first time Neumann BC are imposed by means of an IBM based on a direct heating approach where regularized delta functions are employed. A GCM similar to the one proposed in (Pan, 2010) has also been implemented to validate the numerical procedure so that, after the validation, the Neumann IBM can be finally employed for the simulations of rising adiabatic bubbles in turbulent channel flow. Some of the results reported in this chapter were presented in Santarelli et al. (2014b).

8.2.2 An Immersed Boundary Method for the temperature Dirichlet boundary condition

In the framework of the IBM, (8.1) is modified to account for the disperse phase that is now introduced by means of a heat source q as

$$\frac{\partial T}{\partial t} + \mathbf{u} \cdot \nabla T = a\nabla^2 T + q \tag{8.6}$$

For convenience, the superscript k associated to the Runge-Kutta step is omitted, since the following algorithm applies equally for $k = 1, 2, 3$. For each k, a preliminary temperature field is evaluated without additional heat source as in (8.4) which is now repeated for convenience

$$\begin{aligned}\tilde{T}^k = T^{k-1} &-\gamma_k \, \Delta t \, \nabla \cdot (\mathbf{u}T)^{k-1} \\ &-\zeta_k \, \Delta t \, \nabla \cdot (\mathbf{u}T)^{k-2} + (2\alpha_k)\Delta t \, a\nabla^2 T^{k-1}\end{aligned} \tag{8.7}$$

Afterward the Eulerian temperature is interpolated on the Lagrangian heating points by means of the discrete delta functions δ_h proposed in (Roma et al., 1999)

$$\tilde{T}_L^k(\boldsymbol{X}_L) = \sum_{\boldsymbol{X}_L \in \Omega} \tilde{T}^k(\boldsymbol{x}) \, \delta_h(\boldsymbol{X}_L - \boldsymbol{x}) \, \Delta^3 \, . \tag{8.8}$$

Here, $\boldsymbol{X}_L$ is the position of the Lagrangian heating point and $\boldsymbol{x}$ the position of the Eulerian points involved in the interpolation process, as depicted in Fig. 8.3. Afterward, the Lagrangian heating is evaluated as the difference between the interpolated temperature $\tilde{T}_L^k$ and the desired temperature T_Ω

$$Q^k(\boldsymbol{X}_L) = -\frac{\tilde{T}_L^k(\boldsymbol{X}_L) - T_\Omega(\boldsymbol{X}_L)}{\Delta t} \tag{8.9}$$

and, successively, such values are spread on the Eulerian points by means of the same discrete delta functions

$$q^k(\boldsymbol{x}) = \sum_{l=1}^{N_l} Q^k(\boldsymbol{X}_L) \, \delta_h(\boldsymbol{X}_L - \boldsymbol{x}) \, \Delta V_l \, . \tag{8.10}$$

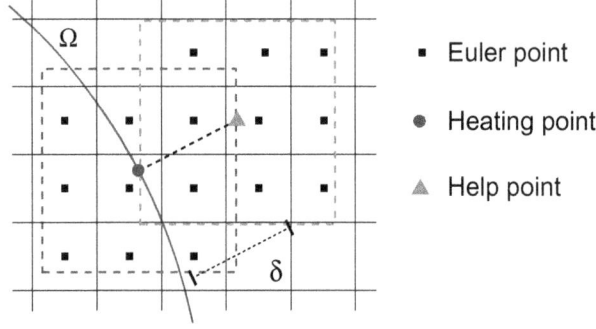

Figure 8.3: Points involved in the definition of the boundary condition for IBM. The dashed squares identify the support of the discrete delta functions for the heating points and for the help points (HP), $\boldsymbol{X}_L$ and $\boldsymbol{X}_H$, respectively.

where ΔV_l is the volume associated to each heating point. Finally, the Eulerian heat source can be introduced in the implicit part of the temperature equation, i.e. (8.5), yielding the final temperature field as

$$
\left[-\frac{T^k}{(\alpha_k)\, a \Delta t} + \nabla^2 T^k \right] = -\frac{1}{a\,(\alpha_k)} \left(\frac{\tilde{T}^k}{\Delta t} + q^k \right) + \nabla^2 T^{k-1} \, . \tag{8.11}
$$

Influence of additional loops

It has been previously mentioned that, in the framework of the IBM, interpolation and successive spreading of field quantities such as velocity and temperature do not yield identical values, as extensively described in Kempe and Fröhlich (2012a) regarding the coupling of the velocity field. For the present study, this implies that after the solution of (8.11), an additional interpolation of the Eulerian temperature on the Lagrangian heating points would not yield the desired values exactly, i.e. $\tilde{T}_L^k(\boldsymbol{X}_L) \neq T_\Omega(\boldsymbol{X}_L)$. Similarly to Wang et al. (2009) and Tavassoli et al. (2013), additional heating loops are hence introduced after (8.11) as follows:

1. The temperature is interpolated on the heating point, yielding $T_L^{(m)}(\boldsymbol{X}_L)$, where m is the loop index. (The superscript k is neglected for clarity.)

2. The Lagrangian heating $Q^{(m)}(\boldsymbol{X}_L)$ is evaluated as as the difference between $T_L^{(m)}(\boldsymbol{X}_L)$ and $T_\Omega(\boldsymbol{X}_L)$.

3. The Eulerian heat source $q^{(m)}(\boldsymbol{x})$ is evaluated and used to correct the temperature field according to
$$
T^{(m)}(\boldsymbol{x}) = T^{(m-1)}(\boldsymbol{x}) + q^{(m)}(\boldsymbol{x})\Delta t \, . \tag{8.12}
$$

In numerical tests it turned out that the correction is small enough so that a repeated evaluation of the other terms of (8.7) and (8.11) is not performed in the same Runge-Kutta sub-step.

The influence of the additional loops is quantified for a simple two-dimensional case, addressing the flow around a cylinder at constant surface temperature T_Ω (by means of a Dirichelet

BC). Based on the inflow velocity and on the diameter, the Reynolds number is $Re_{p,c} = 20$ and the fluid Prandtl number is $Pr = 0.73$. The ratio between the cylinder diameter and the mesh step size is 20, and this results in 62 heating points on the cylinder surface. The CFL number is around 0.7 based on the inflow velocity and on the constant time step. The domain employed is the one used by Wang et al. (2009) for the same problem. Here, the focus is given on the improvement provided by the additional heating loops. The local error on the phase boundary, defined as

$$\epsilon(\boldsymbol{X}_L) = \frac{T_L(\boldsymbol{X}_L) - T_\Omega(\boldsymbol{X}_L)}{T_\Omega(\boldsymbol{X}_L)} , \qquad (8.13)$$

is portrayed in Fig. 8.4 as a function of the angle ϕ_g, where $\phi_g = 0$ is at the front stagnation point. Additionally, the L_1-error and L_2-error, defined as follows, are collected in Tab. 8.1:

$$\epsilon_\infty = \max\left[\epsilon(\boldsymbol{X}_L)\right] \qquad (8.14)$$

$$\epsilon_2 = \sqrt{\frac{\sum_{i=1}^{N_L} \epsilon^2(\boldsymbol{X}_L)}{N_L}} \qquad (8.15)$$

Quantity	$m = 0$	$m = 1$	$m = 3$	$m = 6$
$\epsilon_\infty/10^{-3}$	5.23	3.46	2.87	2.78
$\epsilon_2/10^{-3}$	3.07	2.04	1.67	1.60

Table 8.1: Error on the phase boundary according to (8.14) and (8.15) for different simulations performed, i.e. with a different number m of the additional forcing loops.

Both the curves in Fig. 8.4 and the numerical values in Tab. 8.1 confirm that an adequate number of additional heating loops improves remarkably the performances of the IBM, reducing the local error on the phase boundary and allowing a much better imposition of the desired BC.

Neumann boundary condition

To account for a Neumann BC on the phase boundary, i.e. to apply a given heat-flux K_Ω, the algorithm is slightly modified with respect to the one for a Dirichlet BC. To this end, an additional layer of Lagrangian points, called help points and represented by the symbol $\boldsymbol{X}_H$, is introduced at a distance δ normal to the phase boundary Ω, as portrayed in Fig. 8.3. The method described in Sec. 8.2.2 for a Dirichlet condition is hence modified as follows. First, the temperature $T_H(\boldsymbol{X}_H)$ on the help points and the temperature $T_L(\boldsymbol{X}_L)$ on the heating points are evaluated via Dirac interpolation, as in (8.8). Afterwards, the desired temperature on the phase boundary $T_\Omega(\boldsymbol{X}_L)$ is evaluated by a first order approximation of the temperature gradient in radial direction i.e.

$$T_\Omega(\boldsymbol{X}_L) = T_H(\boldsymbol{X}_H) - K_\Omega \, \delta . \qquad (8.16)$$

Once the desired temperature T_Ω on the phase boundary is obtained, it can be employed in (8.9) so that no additional modification is needed with respect to the procedure for a Dirichlet BC. The parameter δ is set equal to the mesh width Δ, which is identical in all Cartesian directions. This approach is inspired by Zhang et al. (2008) and Pan (2010), as reported in Sec. 8.3.1.

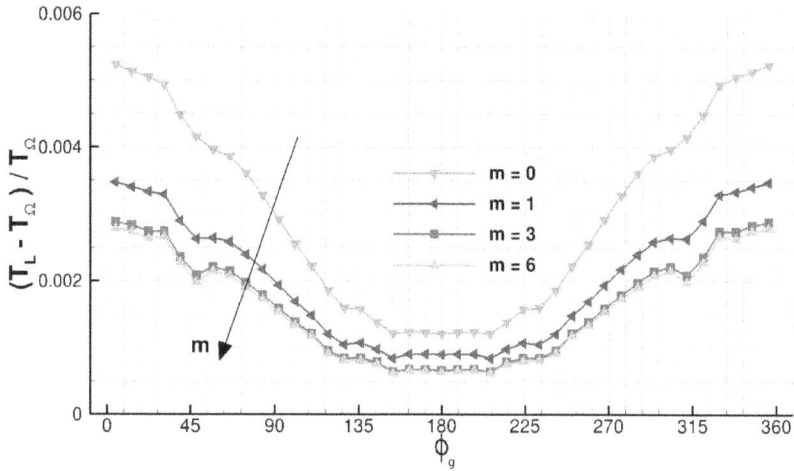

Figure 8.4: Local error according to (8.13) for different simulations where a different number of additional heating loops m was employed. The angle ϕ_g is defined along the cylinder surface and $\phi_g = 0$ at the front stagnation point. Symbols do not coincide with heating points.

8.2.3   The Ghost Cell Method of Pan (2010)

Dirichlet BC To provide a sound validation of the IBM, a Ghost Cell Method (GCM) inspired by Pan (2010) is implemented in the code PRIME and extended to three-dimensional problems. First, a number of Eulerian points, so-called ghost points $\boldsymbol{x}_G$, are detected inside the body, fulfilling the criterion that their normal distance from Ω is smaller than the mesh step size Δ, as portrayed in Fig. 8.5. Then, for each ghost point, two additional points are defined on the straight line normal to Ω passing through the ghost point: A projection point $\boldsymbol{X}_P$ located on Ω, and an image point $\boldsymbol{X}_I$ at a distance δ from Ω. The distance δ is equal $\sqrt{3}h$ so that no ghost point contributes to the computation of the temperature T_I at the image points and it is evaluated, in the three-dimensional domain, from the eight surrounding points by tri-linear interpolation. Then, the temperature on the ghost point

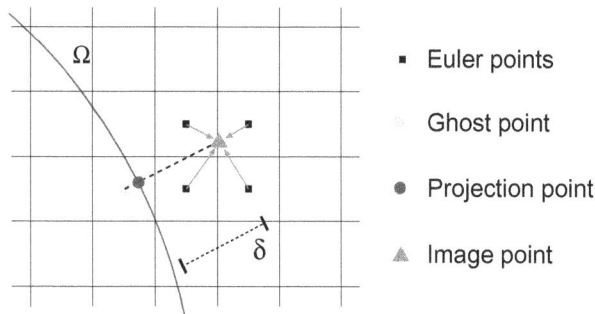

Figure 8.5: Points involved in the definition of the boundary condition for GCM.

is linearly extrapolated from the temperature on the image point and the projection point according to

$$T_G(\boldsymbol{x}_G) = T_P(\boldsymbol{X}_P) - \frac{T_I(\boldsymbol{X}_I) - T_P(\boldsymbol{X}_P)}{\delta} |\boldsymbol{x}_G - \boldsymbol{X}_P| \; . \tag{8.17}$$

The temperature on the ghost points is overwritten by the value from (8.17) after the evaluation of the intermediate temperature (8.7) and after the solution of the Helmholtz equation (8.11). Note that in the CGM the heat source $q(\boldsymbol{x}_{i,j,k})$ in (8.11) is set equal to 0 during all computation steps and the time-advection scheme and the treatment of the diffusive terms is not modified with respect to ones employed in the IBM.

Neumann BC To impose a Neumann BC in the framework of the CGM, the temperature on the ghost point $T_G(\boldsymbol{x}_G)$ is determined by the temperature on the image point and by the temperature gradient K_Ω on Ω according to

$$T_G(\boldsymbol{x}_G) = T_I(\boldsymbol{X}_I) - K_\Omega |\boldsymbol{x}_G - \boldsymbol{X}_I| \tag{8.18}$$

where the same linear approximation of the temperature profile in the radial direction as in the IBM is employed.

8.2.4 Heat transfer around a fixed sphere in cross flow

The IBM and the GCM are validated by simulating the flow around a fixed sphere under different thermal conditions. To this end, three different problems are investigated and described here. *Case*1 deals with the flow field around an isothermal sphere, and an inhomogeneous Dirichlet BC is applied on the phase boundary. In *Case*2, instead, an adiabatic sphere is placed in a temperature field that varies linearly along the y-direction: To this end, a homogeneous Neumann BC is applied on Ω. In the third problem, labeled *Case*3, a sphere with constant heat-flux, imposed by means of an inhomogeneous Neumann BC, is placed in a flow field with otherwise constant temperature, similar to *Case*1. The different thermal BC for the three problems are collected in Table 8.2.

The same domain is used for all simulations presented in this section, consisting of a cubic box of extensions $L \times L \times L$ discretized by 128^3 mesh points. A fixed sphere of diameter $d_p = 0.24H$ is located at $(L/2, L/2, L/2)$ and the ratio between the sphere diameter and the mesh step size is 30. An inflow BC with constant velocity $\mathbf{u} = (u_c, 0, 0)$ and temperature $T = 0$ is applied at the inlet boundary. At the outlet, $\mathbf{u} = (u_c, 0, 0)$ is imposed and a convective BC is applied for the temperature. Homogeneous Neumann BC are applied at the other four boundaries for the velocity field and for the temperature at $z = 0, L$. On the other two boundaries, i.e. $y = 0, L$, the BC for the temperature depend on the problem investigated, as collected in Table 8.2. On the sphere, zero velocity is imposed by means of the IBM of Kempe and Fröhlich (2012a). The Reynolds number based on the inflow velocity and on the diameter is $Re_{p,c} = (u_c \, d_p)/\nu = 10$ and the Prandtl number $Pr = \nu/a$ is 0.71. The CFL number, defined as $(u_c \, \Delta t)/\Delta$ is around 0.74.

The simulations performed with the IBM and the GCM are compared with the corresponding ones performed with the second-order finite volume code ANSYS Fluent using a body-fitted mesh of around 4.7 Million cells. This mesh is fine enough, so that further refinement does not change the solution. For all simulations considered the flow is stationary due to the limited Reynolds number. A two-dimensional representation of the velocity field can be appreciated in Fig. 8.6. Very good agreement for the velocity field is found between the results

Location	Case1	Case2	Case3
Sphere boundary	$T_\Omega = T_{ref}$	$K_\Omega = 0$	$K_\Omega = T_{ref}/a$
$y/L = 0$	$\partial T/\partial y = 0$	$T = 0$	$\partial T/\partial y = 0$
$y/L = 1$	$\partial T/\partial y = 0$	$T = 2T_{ref}$	$\partial T/\partial y = 0$

Table 8.2: Thermal boundary conditions for the three different cases.

on the body fitted mesh and the IBM as shown in Fig. 8.7. For the simulations performed with the IBM and the CGM, the velocity field does not change among the different simulations, since in the present work the temperature field does not influence the velocity field. For all cases the same set of simulation was performed: The simulation with body-fitted mesh to be used as reference, a simulation with the GCM and five simulations with the IBM using different numbers of heating loops $m = 0, 1, 3, 6, 10$.

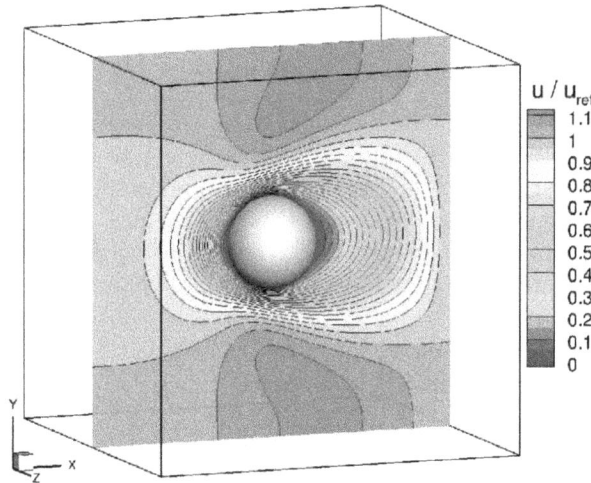

Figure 8.6: Velocity field on the plane $z/L = L/2$.

Case1 The first problem investigated, labeled *Case1*, is the flow around an isothermal sphere where an inhomogeneous Dirichlet condition is applied on the phase boundary, namely $T_\Omega = T_{ref}$. For the simulation performed with the IBM and $m = 3$, the contour plot of the temperature field on the plane $z = L/2$ is portrayed in Fig. 8.8. Figure 8.9 shows the temperature profiles along the center line $(y, z) = (L/2, L/2)$ for the different simulations performed. The temperature evaluated with the GCM provides the larger discrepancy when compared with the reference solution: It is very close to the reference in the flow region at large distance from the sphere but about 10% lower in the stagnation point. With the IBM, the deviation from the reference solution is smaller and the improvement achieved by the additional heating loops can be appreciated, in particular in the zoomed region which represents the region close to the front stagnation point. The simulation with $m = 10$ is almost indistinguishable from the one with $m = 6$ and is therefore not shown. The present results

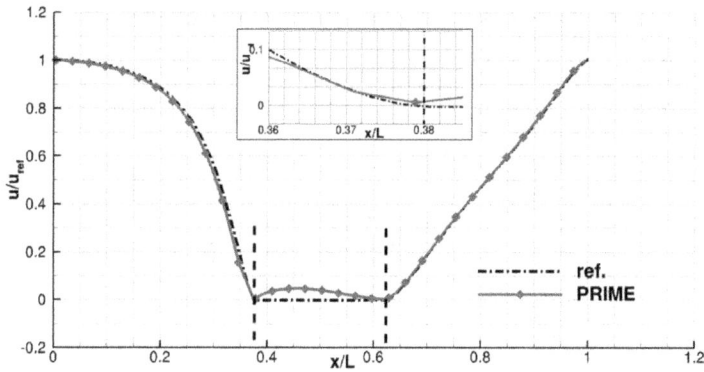

Figure 8.7: Velocity profile along the center line $(y/L, z/L) = (L/2, L/2)$: Comparison between the code PRIME and the reference solution. Zoomed region near the front stagnation point.

show that the IBM performs better than the GCM in this case and that the agreement with the reference solution firts increases with increasing heating loops (up to $m = 1$) and that it is slighlty further off.

Case2 In the second problem an adiabatic sphere, i.e. with zero heat-flux across the sphere surface, is placed in a temperature field that varies linearly across the y-direction: On the bottom wall at $y = 0$, $T = 0$ is imposed and, on the top wall at $y = L$, $T = 2T_{ref}$ is imposed (see Table 8.2). The temperature contour plot on the plane $z = (L/2)$ is portrayed in Fig. 8.10 for the simulation performed with the IBM and $m = 3$. The temperature profiles along the line $(x, z) = (L/2, L/2)$ for the different simulations are presented in Fig. 8.11. With the GCM, the temperature at the front stagnation point is very close to the reference solution, about 0.3% higher, providing a very good agreement. The IBM solution without heating loops is slightly further off, about 5% and improves with increasing number of heating loops. As can be appreciated in the zoomed region, with $m = 3$ the temperature is around 0.5% lower than the reference solution. For a Neumann BC, hence, the GCM performs slightly better than the IBM and the latter strongly improves with additional heating loops.

Case3 The last problem investigated is the flow around a sphere with constant heat-flux imposed on the phase boundary by means of an inhomogeneous Neumann boundary condition on the surface Ω. The BC imposed on the box boundaries are the same as for *Case1*. An impression of the temperature field is provided in Fig. 8.12 by means of the contour plot on the $z/L = L/2$-plane while temperature profiles along the $(y/L, z/L) = (L/2, L/2)$-axis are portrayed in Fig. 8.13. The GCM provides a very good agreement with the reference solution in the outer field. The IBM, as for *Case2*, provides poor agreement with the reference when a limited number of additional loops is employed but improves drastically with m larger or equal 3, eventually reaching the performance of the GCM when compared with the reference solution.

In conclusion, the analysis of three different heat transfer problems related to the uniform flow around a fixed sphere has proved that the GCM and the IBM provide fairly equivalent results for all configurations investigated, matching with a reference solution evaluated on a body-fitted mesh. Furthermore, it is confirmed that, for the IBM, additional heating loops improve the performances of the algorithm, eventually catching up with the results provided by the GCM. Hence, due to the better accuracy expected when moving objects are considered, the IBM is employed to simulate the flow of a swarm of adiabatic bubbles in an turbulent channel flow, as reported in the following section.

Besides the results presented here, both methods were also validated with the analytic solution of a diffusion problem between two concentric cylinders. Different meshes were employed and the rate of convergence was investigated based on the L_2-error determined with respect to the reference. Preliminary results of the convergence analysis for this case may be summarized as follows. When a Dirichlet BC is applied at the inner cylinder, the order of the IBM is around 1 and around 2 for the GCM, as described in Titscher (2012). For a Neumann BC at the inner cylinder the order is around 1.7 for the IBM and around 1 for the GCM, as reported in Baum (2014). Such discrepancies are currently investigated and ongoing research effort is devoted to the definition of further test cases to assess the rate of convergence of both methods. Nevertheless, the author would like to express his gratitude to his both coworkers for the excellent work accomplished.

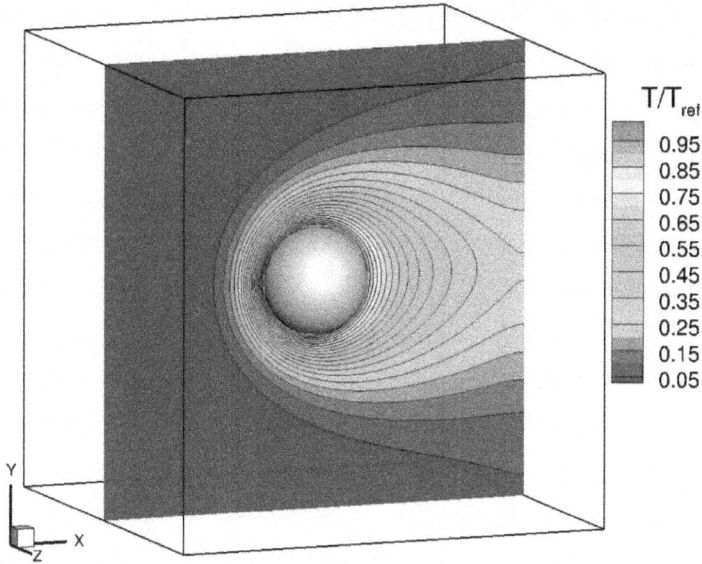

Figure 8.8: Temperature field for *Case1* on the plane $z = L/2$, evaluated with the IBM with $m = 3$.

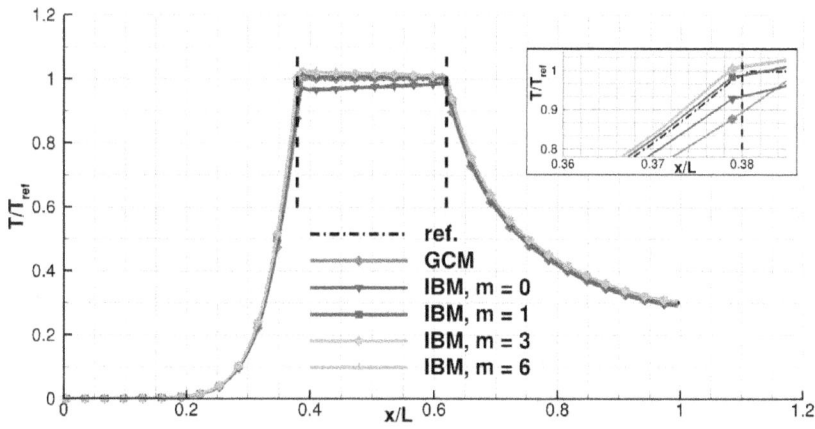

Figure 8.9: Temperature profile along the line $(y, z) = (L/2, L/2)$ for *Case1*: Comparison between the GCM, the IBM with different loops m and the reference solution. Zoomed region near the front stagnation point.

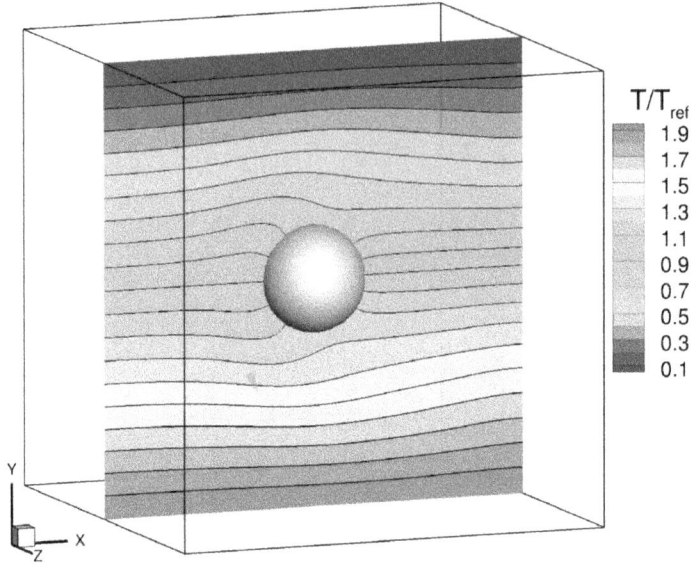

Figure 8.10: Temperature field for *Case2* on the plane $z = L/2$, evaluated with the IBM with $m = 3$.

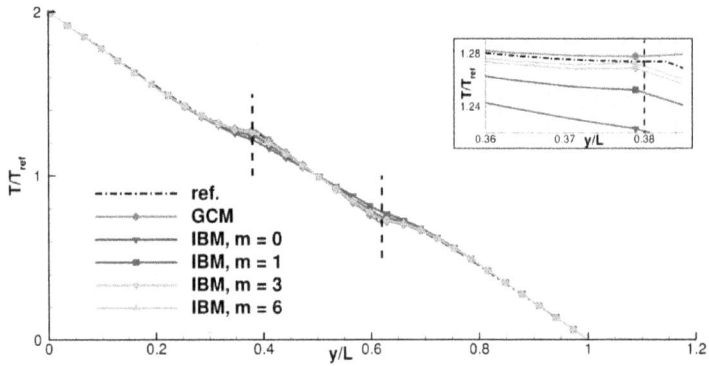

Figure 8.11: Temperature profile along the line $(x, z) = (L/2, L/2)$ for *Case2*: comparison between the GCM, the IBM with different loops m and the reference solution. Zoomed region near the front stagnation point.

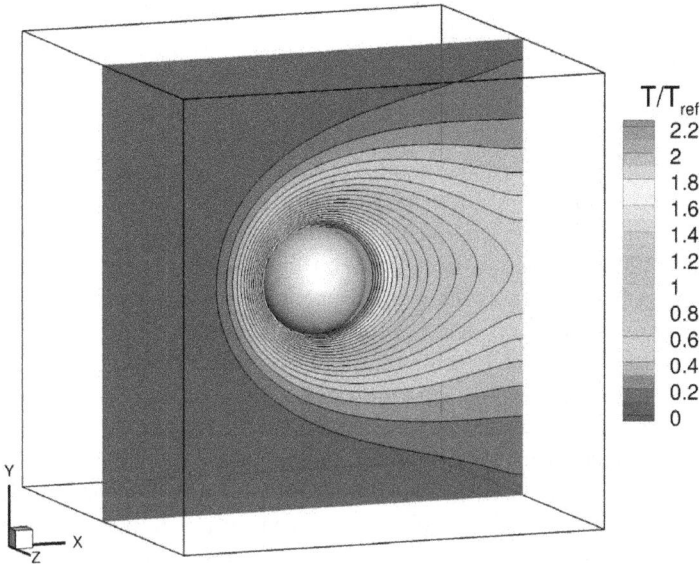

Figure 8.12: Temperature field for *Case3* on the plane $z = L/2$, evaluated with the IBM with $m = 3$.

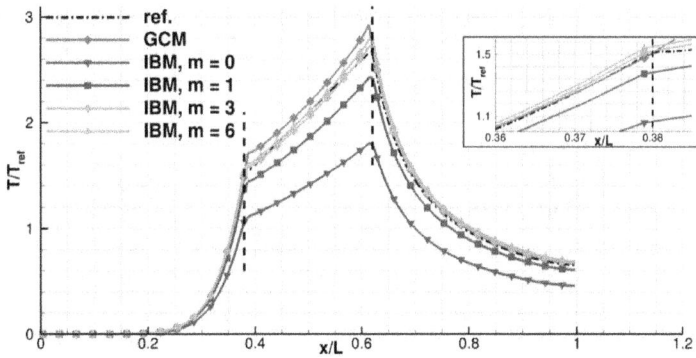

Figure 8.13: Temperature profile along the line $(y, z) = (L/2, L/2)$ for *Case3*: Comparison between the GCM, the IBM with different loops m and the reference solution. Zoomed region near the front stagnation point.

8.3   Heat transfer in bubble-laden turbulent channel flow

8.3.1   Introduction

As for the investigation of the hydrodynamics of bubbles, the channel flow configuration has been often employed to investigate the heat transfer between two-phase mixtures and rigid surfaces due to its resemblance with several industrial plants. Hagiwara (2011) reviewed experimental and numerical studies dealing with this topic, focusing on existing correlations mostly for heat transfer problems where small particles are involved. As stated therein, detailed simulations are needed to provide deeper insight into such phenomena and to improve the existing correlations. Recently, some studies have been presented where such issues are addressed are here reported here focusing on the results regarding the heat transfer between the mixture and the walls. Tanaka et al. (2010) and Tanaka (2011) simulated an upward directed channel flow laden with slightly deformable bubbles employing a VOF approach. It was observed that in general the heat transfer at the walls was increased with respect to the unladen flow due to the larger shear velocity induced by the bubbles. Several simulations were performed with different thermal properties of the mixture. In one of these, the thermal conductivity inside the bubbles was set to a very low value to investigate the insulating effect of the bubbles, but a limited influence of this parameters was found on the overall heat transfer. Other simulations allowed observing that the heat transfer enhancement is higher when the Prandtl number of the fluid is larger. With respect to an unladen flow, the highest increase of the wall heat transfer, quantified by the Nusselt number, was around 30% for $Pr = 1$. Dabiri and Tryggvason (2013, 2015) investigated the same configuration, i.e. a turbulent channel flow laden with bubbles. As in (Tanaka, 2011), nearly spherical bubbles rising prevalently near the walls enhance the heat flux, due to the agitation of the boundary layer induced by the bubbles. With respect to an unladen flow, the highest increase of the heat transfer of around 60% was found for nearly spherical bubbles at moderate Eötvös number and very large Prandtl number, $Eo = 0.9$ and $Pr = 7$, respectively. Deen and Kuipers (2013), in a somewhat simpler configuration, investigated the heat transfer between a hot wall and the two-phase mixture, the latter consisting of quiescent fluid and rising bubbles. For a single bubble rising close to a vertical hot wall, these authors observed an increase in the local heat transfer in correspondence of the the bubble center, due to the thinner boundary layer induced by the presence of the bubble. With increasing vertical distances (in the downward direction) from the the point where it exhibits its maximum, the heat transfer decreases due to the downward motion of hot fluid between the wall and the bubble, which enlarges the thermal boundary layer. Further away from the bubbles the heat transfer exhibits an other local maximum.

To address these problems and to contribute to a better understanding of the phenomena involved, three simulations are performed to investigate the influence of bubbles on the heat transfer between the two-phase mixture and the walls. The results of this analysis are reported in the following section where the focus is given on the heat transfer between the mixture and the wall, as in the studies mentioned above, since this quantity is crucial for many industrial application.

8.3.2 Numerical setup

Several phenomena may play a role for the heat transfer in two-phase flows, such as phase change, nucleation of bubbles and evaporation/condensation of the fluid. To reduce the investigation field and focus only on the selected mechanisms, such phenomena are not considered here. Here, with respect to single-phase case, two main mechanisms are expected to influence the heat transfer in the bubbly flow: The modification of the convective heat transfer due to the presence of the bubbles and their insulating effect. To distinguish between these two effects, three simulations will be analyzed:

- A single-phase flow simulation labeled *Unladen*. It it the same simulation introduced in Sec. 8.1.2 and is here used as reference.

- A simulation with bubbles where no thermal boundary condition is applied on the phase boundary. This corresponds to a mixture of two fluids with the same thermal properties and implies that the temperature field is influenced only by bubble dynamics. This simulation is the one labeled *SmMany* in Chap. 3 where, in addition, a transport equation for the temperature without heat sources is solved.

- A simulation with the same bubbles as *SmMany* where adiabatic bubbles are considered by means of a homogeneous Neumann condition on the phase boundary. This simulation is labeled *Adia* and allows to address the influence of the insulating effect of bubbles on the heat transfer.

Since the temperature field has no influence on the velocity and on the pressure field nor on the bubble dynamics, case *SmMany* and *Adia* are statistically identical from a hydro-dynamical point of view: Fluid statistics regarding the velocity field and bubble statistics, discussed in Sec. 3.3, are, hence, statistically equivalent among the two simulations. The only difference between *SmMany* and *Adia* is the thermal boundary condition applied on the bubble surface and, therefore, the temperature field.

For all simulations, the vertical walls are kept at a fixed temperature, namely $T(x, z) = T_{ref}$ for $y/H = 0$ and $T(x, z) = -T_{ref}$ for $y/H = 1$, while periodic BC are applied in the streamwise and in the spanwise direction for the temperature field as for the velocity and pressure fields. The fluid Prandtl number is 1.

To perform the *Adia* simulation, i.e. to apply a homogeneous Neumann BC on the phase boundary for the temperature, the IBM method proposed in Sec. 8.2.2 was employed. As in the case of a fixed sphere, the distance δ between the Lagrangian $\boldsymbol{X}_L$ and the help points $\boldsymbol{X}_H$ was set equal the step size of the mesh Δ. The number of additional heating loops is $m = 3$, which was proved to be a good compromise between performances and computational resources. The algorithm was slightly modified to account for bubble-bubble and bubble-wall interaction since the stencil of the help point $\boldsymbol{X}_H$, where interpolation and spreading take place, can be quite distant from the phase boundary. In the worst case, the farthest point involved can be three mesh steps away from the phase boundary. When a bubble approaches the wall, it happens that some points of the stencil lie outside the physical domain. The heating process is performed only for Lagrangian points whose stencil presents at most one point outside the physical domain. For the Lagrangian points whose stencil presents two points outside the physical domain, the heating process is not performed since there is no second point outside the physical domain to perform the interpolation. A similar criterion is employed to account for two bubbles approaching each other, according to which the heating process for the Lagrangian points of one bubble is performed only if at most one point of the stencil is inside the other bubble. This decision was taken since inaccuracies can arise when a large portion of the stencil employes temperature values inside a bubble. Nevertheless, the stencils of two bubble can share some Eulerian points, as long as the aforementioned criterion is satisfied.

8.3.3   Flow visualizations

The role of the Neumann BC on the temperature field is first investigated for the local field around a rising bubble close to the wall. Figure 8.14, left, shows the instantaneous temperature field in the vicinity of a bubble near the wall for the *SmMany* case on the vertical plane cutting the bubble center. As expected, the temperature field in the region above the bubble is almost not influenced by the bubble presence. From the point where wall and bubble surface are closest, the iso-lines of the temperature are influenced by the velocity field induced by the bubble, and are convoyed toward the rear point of the bubble. This shows that the temperature field is modified by the bubbly only due to its dynamics, i.e. due to the velocity field induced by the bubble. Figure 8.14, right, shows the instantaneous temperature field in the vicinity of a bubble near the wall for the *Adia* case and the scenario is quite different with respect to the *SmMany* case. The influence of the bubble is revealed both in the the upper and in the lower region and iso-lines of the temperature impinge normally to the phase boundary, as expected when a homogeneous Neumann condition is applied. In order to gain a first impression of the role of bubbles into the heat transfer at

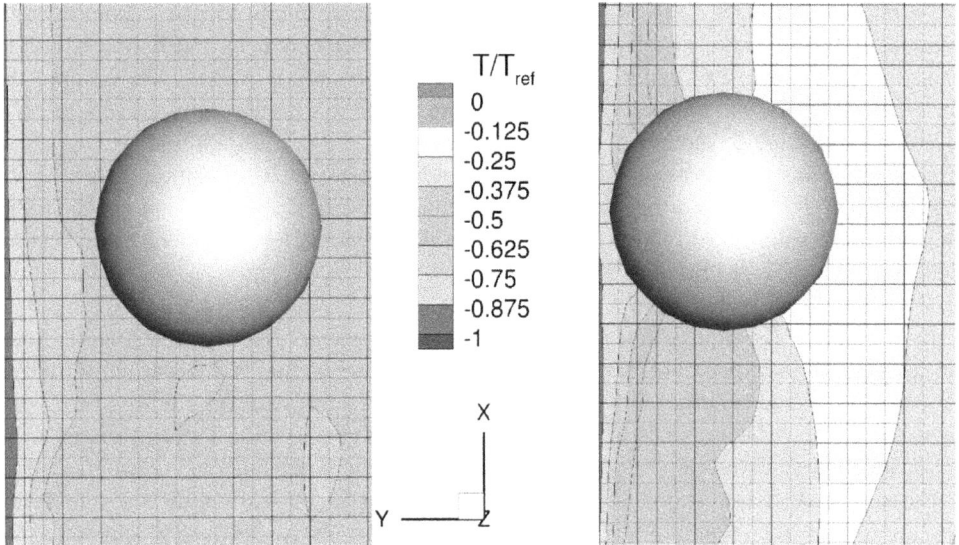

Figure 8.14: Temperature field in a vertical z-plane cutting the bubble center in the vicinity of the wall for simulation *SmMany* (left) and *Adia* (right).

the wall, Fig. 8.15 portrays the instantaneous fluctuations of the fluid streamwise velocity (left) and the temperature value (right) on the $y/H = 0.002$-plane for the *Unladen* case. The following considerations can be drawn for the hot wall at $y/H = 0$ but apply also for the cold wall at $y/H = 1$. Instantaneous fluctuations in a y_w/H-plane are defined as follows

$$u'(x, y_w, z, t) = u(x, y_w, z, t) - \langle u(y_w) \rangle_{xzt} \tag{8.19}$$

and normalized with the bulk velocity U_b. Regarding the fluctuations, the streaky structures usual in unladen channel flow can be observed, both the ones consisting of positive fluctuations and the ones consisting of negative fluctuations. Such structures strongly influence the temperature field and the heat transfer at the wall, as portrayed in Fig.8.15, right. Negative (positive) fluctuation regions correspond to high (low) temperature regions, i.e. of region with lower (higher) heat flux, as observed by Kim and Moin (1989) in the same channel flow configuration. This correlation can be easily explained if one considers that positive fluctuations $u' > 0$ correspond to sweep events that bring fluid with lower temperature towards the wall, increasing the temperature difference, and hence the heat flux. Negative fluctuation regions instead, mainly associated with ejections events, push fluid with high temperature from the wall into the core region, enlarging the zones with high temperature and hence reducing the heat flux.

A similar, though more complex correlation between fluid fluctuations and temperature can be observed also in the flow laden with bubbles in the *SmMany* case, as portrayed in Fig. 8.16, where only bubbles with $y_p + r_p < 0.0022H$ are portrayed. As already discussed in Sec. 3, the presence of the bubbles hamper the usual coherent structures in the wall region but, nevertheless, a local correlation can be observed between negative (positive) fluctuation regions and high (low) temperature zones. As depicted in Fig. 8.16, the positive fluctuation regions correspond to the regions of the bubble wakes: The wakes play the role of the sweeps and agitate the fluid. This means that hot fluid is pushed towards the core region and cold fluid towards the wall, yielding a higher heat flux. Such a straightforward correspondence as in the unladen case is not always observed: There are some high (low) temperature regions which do not correspond to negative (positive) fluctuations and the correlation between fluid and temperature seems much more complicated than in the unladen flow.

For the case *Adia*, portrayed in Fig. 8.17, the picture is quite similar and no substantial differences can be appreciated with respect to the *SmMany* case. Indeed, as discussed previously, the modification of the temperature field due to the zero heat flux condition on the bubble surface is mainly to be found in the vicinity of the bubble, and its influence on the temperature field is apparently much lower compared to the agitation of the fluid due to the bubbles and their wakes.

8.3.4 Statistics of the temperature field

After the instantaneous flows were discussed, some statistical quantities are now presented. As for velocity-related statistics, average is performed over the two periodic directions x and z and over time, when not otherwise stated. Figure 8.18 portrays the time history of the wall heat transfer, where the instantaneous wall heat transfer $\tilde{q}_w(t)$ is defined as follows

$$\tilde{q}_w(t) = -\frac{1}{2} \left(a \frac{\partial T(t)}{\partial y} \bigg|_{y/H=0} + a \frac{\partial T(t)}{\partial y} \bigg|_{y/H=1} \right) \tag{8.20}$$

and its values is normalized with the reference value q_0, defined as $q_0 = a\Delta T/H$. In a laminar unladen channel flow, q_w/q_0 is equal 1. As for the wall shear stress evaluated according to (3.10) in Sec. 3.3.2, due to the collision model there is always at least one Eulerian point

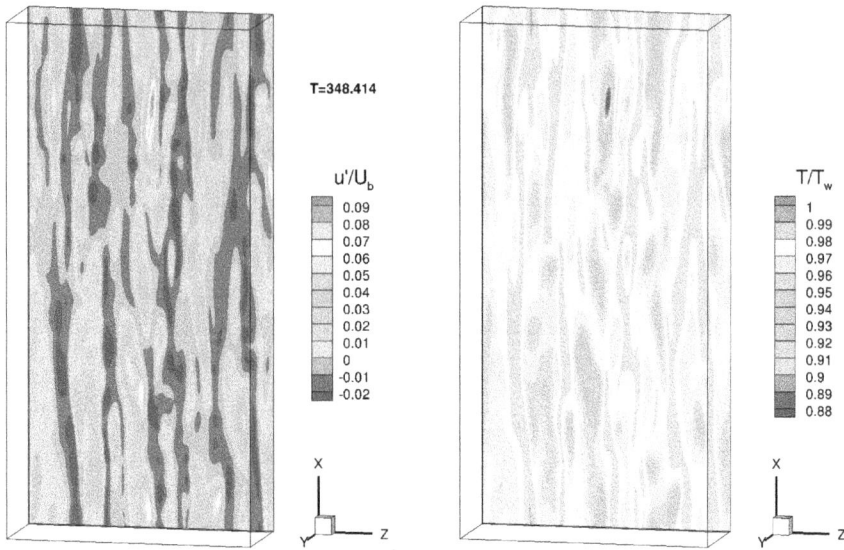

Figure 8.15: Fluctuations of of streamwise fluid velocity according to (8.19) (left) and temperature values (right) on the $y/H = 0.002$ for *Unladen*.

free of bubbles close to the wall for the evaluation of (8.20) (see Fig. 2.1 in Sec. 2.3). As portrayed in Fig. 8.18 the wall heat coefficient of the unladen flow is much lower than the ones of the laden simulations, and the fluctuations around the mean values are quite limited. For the simulations with bubbles the magnitude is quite similar and in both cases larger fluctuations around the mean value are observed. The time averaged values of the heat transfer are collected in Table 8.3. The enhancement of the wall heat transfer, up to 70% higher with respect to the *Unladen* case, is due to the mixing caused by the bubbles and by the turbulence that is enhanced in the wall region since the higher turbulence level reduces the thickness of the boundary layer yielding a larger heat flux. The heat fluxes among the two simulations with bubbles is very similar: Apparently the insulating effect of bubbles induced by the Neumann BC plays a marginal role with respect to the influence of the bubble-induced turbulence for the present parameter range. The slight influence of the thermal BC is in agreement with the results in (Tanaka, 2011) which draw the same conclusion, as previously mentioned. Nevertheless, q_w for the *Adia* case is around 2% lower than in the *SmMany* case, being the lower value related to the local insulating effect of the bubbles, as depicted in Fig. 8.14.

Why is the insulating effect of bubbles on the overall heat transfer so limited? When compared with an unladen flow, two mechanisms contribute to the wall heat transfer: The agitation of fluid induced by the bubble and the modification of the temperature field due to the thermal properties of the bubbles. For the first mechanism, the convective one, two main aspects can be mentioned: The flow of cold fluid between the wall and the bubble and the agitation of the fluid induced by the bubble wake whereby the mutual contribution of such effects depends on many factors. One is the wake structure of the bubble, i.e. the bubble Reynolds number: For small Re_p the wake agitation is small, while for large Re_p the agitation is large and dominant in the convective heat transfer. For the investigated regime,

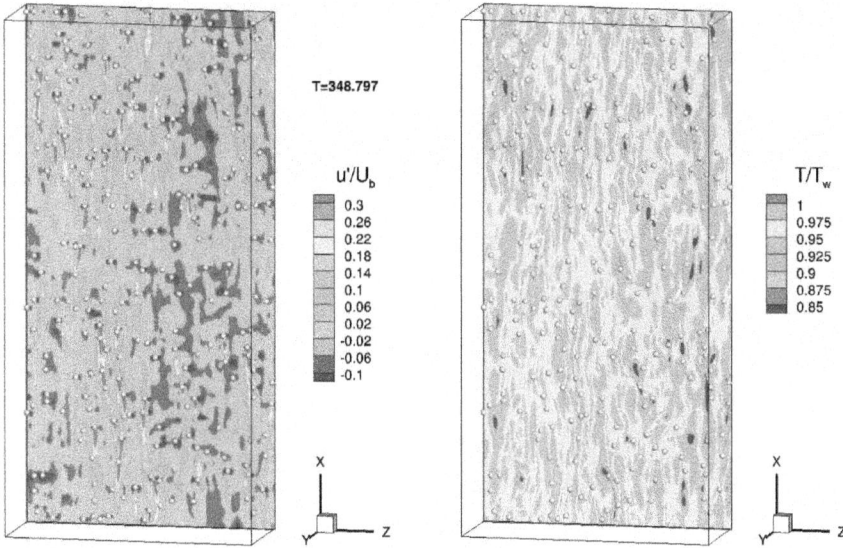

Figure 8.16: Fluctuations of of streamwise fluid velocity according to (8.19) (left) and temperature values (right) on the $y/H = 0.002$ for *SmMany*. Only bubbles with $y_p + r_p < 0.09H$ are represented.

the latter effect is supposed to play the larger role as depicted in Fig. 8.17 where a strong relation is observed between heat transfer and regions where the bubble wake are expected. The conductive mechanism, i.e. the one related to the thermal BC on the phase boundary, is a local effect, since it influences the temperature field only in the vicinity of the bubble, as portrayed in Fig. 8.14. For the chosen regime, the agitation of the fluid induced by the wake of the bubbles extends to flow regions much larger than the ones where the conductive mechanism is present. Hence, the insulating mechanism due to the bubble thermal properties, plays only a marginal role for the investigated parameter range. Nevertheless, it is supposed to play a larger role for mixtures where the wake-related agitation is reduced, e.g. for lower Re_p, and for configurations where the portion of wall directly influenced by the bubble presence is larger than the present case.

Quantity	*Unladen*	*SmMany*	*Adia*
q_w/q_0	7.12	12.17	11.92

Table 8.3: Time-averaged heat transfer coefficient of the three simulations.

Regarding the enhancement of the wall heat transfer, the following observations can be made for the comparison with similar works in the literature. The augmentation of the heat flux with respect to the single-phase flow is higher than in Tanaka (2011) and the larger enhancement is related to the much higher turbulence augmentation induced by the bubbles in the present study compared to (Tanaka, 2011). The same reason applies to the comparison with results by Dabiri and Tryggvason (2015), who found an enhancement of the wall-heat flux of around 33% for the configuration that mostly resembles the parameter range of the present

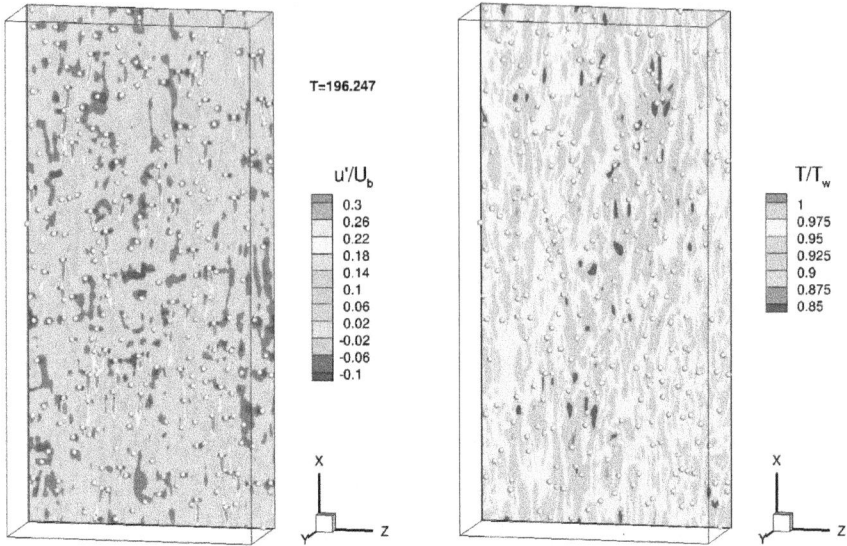

Figure 8.17: Fluctuations of of streamwise fluid velocity according to (8.19) (left) and temperature values (right) on the $y/H = 0.002$ for *Adia*. Only bubbles with $y_p + r_p < 0.09H$ are represented.

study (Case 8 therein).

Temperature-related, turbulent statistics are portrayed in Figs. 8.19-8.20. The averaging time is around $400T_b$ for all three simulations and Eulerian points inside the bubbles are excluded from the averaging process, as for the velocity-related statistics. The mean temperature profile is modified by the presence of the bubbles: It is steeper at the wall for $y_w/H < 0.04$, as expected by the q_w-values present in Table 8.3. This higher inclination is compensated in the core region, where $\partial \langle T \rangle / \partial y$ is somewhat lower than in the unladen case. No large difference can be appreciated between the *SmMany* and the *Adia* case. The temperature fluctuations $\langle T'T' \rangle$ are higher for the simulations with bubbles in the center region, while the maxima in the near-wall region are higher in the unladen case. The maxima for *SmMany* and *Adia* are closer to the wall with respect to the maxima of the *Unladen* case: This is due to the higher friction velocity, as for the maxima of fluid statistics presented in Chap. 3. For the *Adia* case, fluctuations at the wall are slightly higher than in the *Sm-Many* case. The reason for this is the influence of the Neumann BC which is larger where temperature gradients are larger, yielding higher fluctuations. In the center region, instead, fluctuations are higher for the *SmMany* case: Where temperature gradients are lower, the influence of the Neumann BC is reduced.

Turbulent fluxes for the temperature are presented in Fig. 8.19 and the shape of both fluxes resemble the ones of the unladen channel. The streamwise flux $\langle u'T' \rangle$ is lower for the bubble-laden simulations: this mean that the correlation between temperature and streamwise fluid velocity is reduced by the presence of the bubbles. As for the temperature fluctuations, the maxima are shifted towards the wall. The wall-normal flux $\langle v'T' \rangle$, instead, is almost twice as high as in the unladen case but, to this point, no explanation could be found.

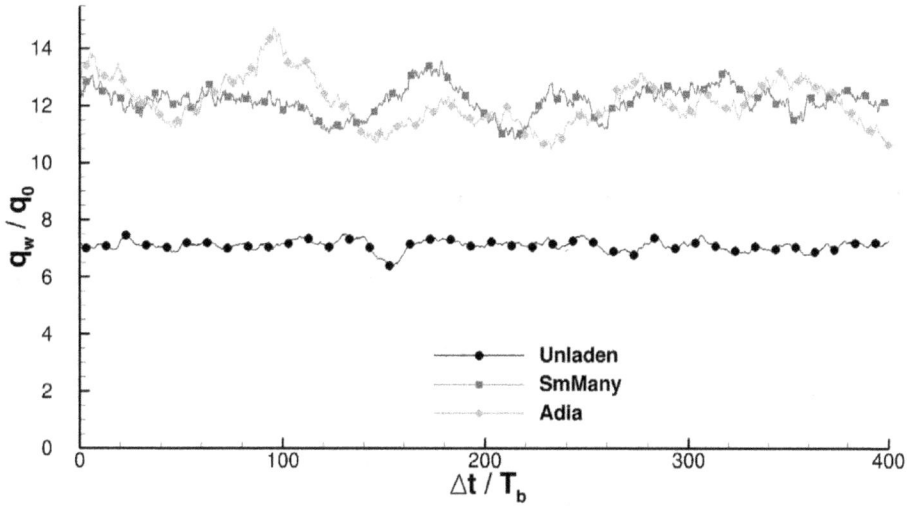

Figure 8.18: Time history of wall heat coefficient q_w/q_0 for the three simulations.

In conclusion, the influence of air bubbles on the wall heat transfer in upward channel flows was investigated by means of instantaneous flow visualizations and statistical quantities. The heat transfer between the walls and the mixture is highly increased by the presence of the bubbles: In the present configuration and for the chosen parameter range, the enhancement is around 70%. This is due to the velocity field induced by the bubbles a the walls, which enhances the turbulence level and reduces the thickness of the boundary layer, yielding an increased heat transfer. A homogeneous Neumann BC on the phase boundary was imposed to account for the thermal properties of the bubbles and their insulating effect. A small reduction of the time-averaged heat transfer was observed when compared with an analogous simulation where no thermal BC is prescribed on the phase boundary. Nevertheless, this effect plays only a marginal role compared to bubble-related turbulence enhancement for the chosen parameter range. The insulating effect of air bubbles is expected to play a larger role in other configurations where the turbulence induced by the bubbles is not so strong, as discussed above. As for the fluid and bubble statistics presented in the previous chapters, the results regarding the temperature field can now be used for validation and for the development of sophisticated models for non-isothermal bubbly flows.

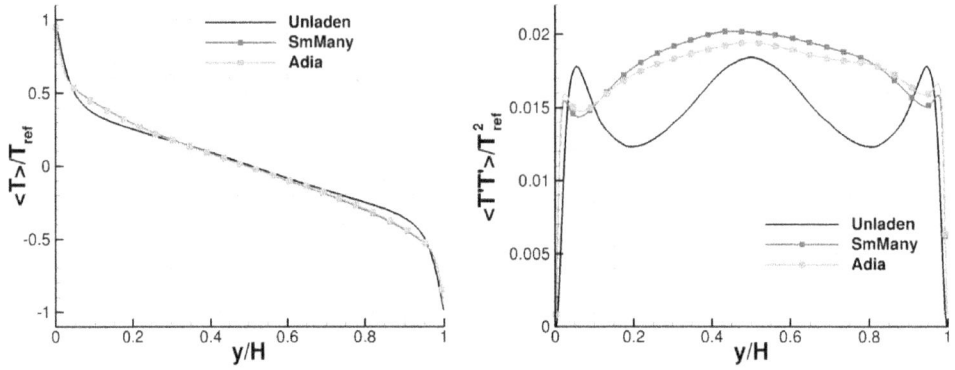

Figure 8.19: Mean temperature profile (left) and mean temperature fluctuations for the three simulations.

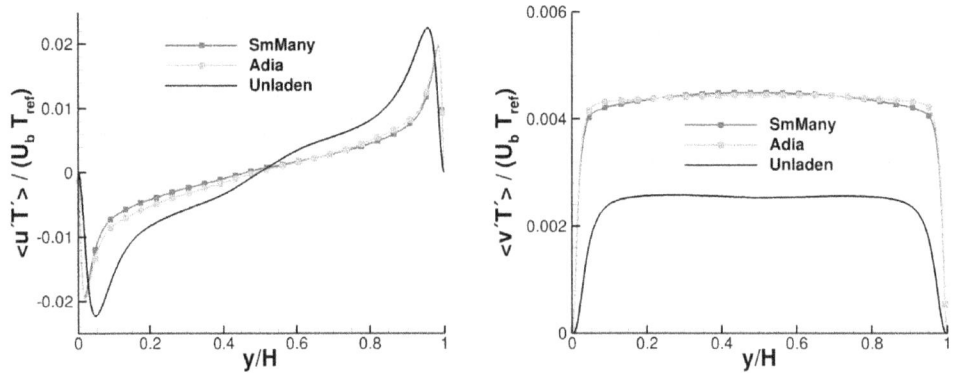

Figure 8.20: Turbulent temperature fluxes: $\langle u'T' \rangle$ (left) and $\langle v'T' \rangle$ (right).

9   Summary and outlook

Direct Numerical Simulations are performed to investigate the dynamics of spherical bubbles in turbulent channel flow. Bubbles are introduced in the Navier-Stokes equations by means of additional forces by means of an Immersed Boundary Method (IBM). The bubbles are represented by spheres of rigid shape and a no-slip condition is applied at the phase boundary between fluid and bubble velocity, which matches the physics of small bubbles rising in contaminated water. Quantitative statistical data regarding both phases are provided and can be now employed for the improvement and development of two-phase turbulence modeling.

The numerical method employed is validated for two reduced problems for the chosen regime, using the same spatial discretization employed in the channel simulations. Good agreement regarding the global forces on the bubbles is found with reference data for the rise of a bubble in quiescent fluid and for the response of a fixed bubble in cross-flow with constant shear rate. The validation is performed for small and for larger bubbles.

The analysis of the two-phase mixtures starts with the study of the the flow of a dense swarm of small bubbles which is used as a reference throughout this work. The flow modifications induced by the bubbles with respect to the single-phase flow are quantified by tree- and two-dimensional fluctuations of the fluid which allow observing flows structures that are not found in the unladen case. Similar structures, whose characteristic length is quantified by two-point correlation functions, have been observed in a somewhat similar configuration, i.e. for sedimenting particles in channel flow (Uhlmann, 2008; Garcia-Villalba et al., 2012). An analysis of the turbulence statistics (wall shear stress, velocity profiles, components of the Reynolds stress tensor, etc..) is performed and modifications with respect to the unladen case are put in connection with bubble statistics. For the chosen regime, the turbulence of the liquid phase is increased by the bubbles, in agreement with correlation proposed in the literature. The hindrance effect, according to which a swarm rises slower on average than a single bubble, is quantified and matches qualitatively with experimental results in the literature. In the dense swarm, bubbles align mainly horizontally for small distances and this is due to a pressure-related attraction mechanisms between close bubbles.

The influence of the total void fraction is addressed by the analysis of a simulation where a dilute swarm of small bubbles is investigated. For the chosen regime, the enhancement of the turbulence of the liquid is related to the total void fraction, i.e. it is lower for the dilute swarm. As for the denser swarm, the turbophoresis effect and the shear-induce lift force contribute to the shape of the averaged bubble distribution. For small distances bubbles align mainly vertically and this is related to the aspiration effect induced by the bubble wakes. Due to the larger bubble distance, the pressure-related attraction mechanism plays a marginal role in the dilute swarm.

A swarm consisting of large bubbles with the same total void fraction as for the denser swarm is simulated to investigate the influence of the bubble size. The fluid structures induced by the large bubbles are larger than the corresponding ones induced by the small bubbles, eventually exceeding the dimension of the channel. Turbulence of the liquid phase is strongly increased by the large bubbles and this feature is related to the dynamics of the bubbles themselves. It is shown that the different types of path for bubbles of different size (i.e. different Reynolds number) observed for bubbles rising in quiescent fluid are also observed in the turbulent channel flow. This is portrayed in flow visualizations and quantified by fluctuations of the bubble velocities. The preferential horizontal alignment holds, though reduced, also for large bubbles for which the vertical alignment is the least probable because of the complex paths followed by the large bubbles.

In a bidisperse swarm several flow quantities resemble the ones in the monodisperse swarms. Nevertheless, the large bubbles induce larger velocity fluctuations of the small bubbles that are larger when compared with the monodisperse swarm and, on the other hand, the fluctuations of the large bubbles are reduced with respect to the monodisperse case. A preferential oblique alignment is observed when mixed bubble pairs are considered and this is due to the pressure-related attraction mechanism and to the different rise velocity of the bubbles. Involved quantities as well as flow visualizations are reported to back up these observations.

The influence of the direction of the fluid velocity is investigated by means of a simulation in a downward configuration, where all other parameters are the same employed in the reference case. In the core region the fluid structures in the streamwise direction induced by the bubbles are similar to the ones in the upward flow, while in the near-wall region the fluid structures are larger both in the streamwise and in the spanwise direction. Such wall-bounden structures, observed also in instantaneous flow visualizations, are found only in the downward configuration. The enhancement of the turbulence of the liquid phase is stronger in the downward case and the bubble agitations, quantified by bubble velocity fluctuations, is stronger, in accordance with previous studies in the literature. The spatial distribution of bubbles is more similar to a random one than in the upward case and the preferential horizontal alignment is still observed but reduced.

A detailed analysis of the modification of the turbulence of the fluid induced by the bubbles is provided. For this purpose a budget analysis of the transport equation of the turbulent kinetic energy (TKE) for two-phase flows is performed. For the investigated regime, bubbles act as source terms for the fluctuating flow of the liquid and, with respect to the unladen case, the dissipation rate of the TKE is strongly increased by the bubbles. An analysis of each term of the transport equation of the local instant kinetic energy of the liquid phase is performed to address the local instant modification of the turbulence induced by the bubbles and dedicated flow visualizations are provided to address the different mechanisms involved.

To investigate the heat transfer in bubbly flows, the original IBM is improved to account for the temperature field of two-phase mixtures. To this end the transport equation of the temperature is implemented in the code PRIME and validated. For a correct application of thermal boundary conditions (BC) on the phase boundary two approaches have been considered: A Ghost Cell Method, inspired by Pan (2010), and a *thermal* IBM, proposed by Wang et al. (2009). The latter has been improved to account for a Neumann BC on the phase boundary, since the thermal coupling of air bubbles in water can be represented by a homogeneous Neumann BC. The improved IBM was employed to investigate the thermal behavior of a bubble swarm in an upward turbulent channel flow. With respect to the unladen flow, the presence of the bubbles strongly increases the heat transfer between the mixture and the walls due to the turbulence enhancement induced by the bubbles. For the investigated regime the thermal insulating effect of bubbles plays only a marginal role.

Steven Deeks of the University of California stated in September 2014 that "The best scientific studies raise as many questions as answers"[1] and, although the evaluation of the present work goes beyond the intention of the author, this study actually raises several research questions that future research may consider. Additional simulations of the same configuration with further variation of parameters, for example, would enrich the presented investigation and allow a wider analysis: A swarm at even higher total void fraction, a polydisperse swarm consisting of bubbles of more than two classes, a downward simulation with a different ratio between buoyancy and drag force induced by the fluid, to mention but a few. The deformability of the bubbles is also supposed to play a significant role for certain flow regimes, as reported for example by Dabiri et al. (2013). In the framework of the IBM, Schwarz (2014) developed an algorithm to simulate deformable bubbles, where the bubble surface is analytically described by spherical harmonics functions and where the bubble deformation is related to the pressure field of the surrounding fluid. The simulation of swarms of deformable bubbles would hence allow accounting for a wider physical spectrum of bubbly flows. Regarding the analysis of the turbulence of the liquid phase, an assessment of the existing closure models may be an appealing complement of the present study. If the existing models would prove as unable to correctly represent the turbulence of the liquid, possibly new closure relations may be developed based on the detailed quantitative data provided in the present work.

---

[1]http://news.sciencemag.org/health/2014/09/how-did-berlin-patient-rid-himself-hiv

Bibliography

Adoua, R., Legendre, D., and Magnaudet, J., 2009. Reversal of the lift force on an oblate bubble in a weakly viscous linear shear flow. *Journal of Fluid Mechanics*, 628:23–41.

Aland, S., Schwarz, S., Fröhlich, J., and Voigt, A., 2013. Modeling and numerical approximations for bubbles in liquid metal. *European Physics Journal - Special Topics*, 220: 185–194.

Baehr, H. D. and Stephan, K., 2004. *Wärme- und Stoffübertragung*. Springer.

Bagchi, P. and Balachandar, S., 2002. Effect of free rotation on the motion of a solid sphere in linear shear flow at moderate Re. *Physics of Fluids*, 14(8):2719–2737.

Bagchi, P., Ha, Y. M., and Balachandar, S., 2001. Direct numerical simulation of flow and hear transfer from a sphere in a uniform corss-flow. *Journal of Fluid Engineering*, 123: 347–358.

Bagchi, P. and Balachandar, S., 2004. Response of the wake of an isolated particle to an isotropic turbulent flow. *Journal of Fluid Mechanics*, 518:95–123.

Baum, O., 2014. Implementierung und Validierung verschiedener Immersed-Boundary-Methoden für die Temperatur. Master's thesis, Technische Universität Dresden, Institute of Fluid Mechanics.

Bhagwat, S. M. and Ghajar, A. J., 2012. Similarities and differences in the flow patterns and void fraction in vertical upward and downward two phase flow. *Experimental Thermal and Fluid Science*, 39:213–227.

Bolotnov, I. A., Jansen, K. E., Drew, D. A., Oberai, A. A., Lahey Jr., R. T., and Podowski, M. Z., 2011. Detached direct numerical simulations of turbulent two-phase bubbly channel flow. *International Journal of Multiphase Flow*, 37:647–659.

Bouchet, G., Mebarek, M., and Dusek, J., 2006. Hydrodynamic forces acting on a rigid fixed sphere in early transitional regimes. *European Journal of Mechanics - B/Fluids*, 25(3):321 – 336.

Bröder, D. and Sommerfeld, M., 2002. An advanced LIF-PLV system for analysing the hydrodynamics in a laboratory bubble column at higher void fractions. *Experiments in Fluids*, 33(6):826–837.

Bunner, B. and Tryggvason, G., 2002a. Dynamics of homogeneous bubbly flows. Part 1. Rise velocity and microstructure of the bubbles. *Journal of Fluid Mechanics*, 466:17–52.

Bunner, B. and Tryggvason, G., 2002b. Dynamics of homogeneous bubbly flows. Part 2. Velocity fluctuations. *Journal of Fluid Mechanics*, 466:53–84.

Bunner, B. and Tryggvason, G., 2003. Effect of bubble deformation on the properties of bubbly flows. *Journal of Fluid Mechanics*, 495:77–118.

Cartellier, A. and Riviere, N., 2001. Bubble-induced agitation and microstructure in uniform bubbly flows at small to moderate particle Reynolds numbers. *Physics of Fluids*, 13(8): 2165–2181.

Clift, R., Grace, J., and Weber, M., 2005. *Bubbles, Drops, and Particles*. Dover Civil and Mechanical Engineering Series. Dover Publications.

Crowe, C., 2005. *Multiphase Flow Handbook*. Mechanical and Aerospace Engineering Series. Taylor & Francis.

Dabiri, S. and Tryggvason, G., 2013. Turbulent bubbly channel flow and its effect on heat tranfer. In *Proceedings of the ASME 2013 Fluids Engineering Division Summer Meeting, Incline Village, USA*, number FEDSM2013-16217.

Dabiri, S. and Tryggvason, G., 2015. Heat transfer in turbulent bubbly flow in vertical channels. *Chemical Engineering Science*, 122:106–113.

Dabiri, S., Lu, J., and Tryggvason, G., 2013. Transition between regimes of a vertical channel bubbly upflow due to bubble deformability. *Physics of Fluids*, 25:102110.

Deen, N. G. and Kuipers, J., 2013. Direct numerical simulation of wall-to liquid heat transfer in dispersed gas-liquid two-phase flow using a volume of fluid approach. *Chemical Engineering Science*, 102:268–282.

Deen, N. G., Van Sint Annaland, M., and Kuipers, J., 2004. Multi-scale modeling of dispersed gas-liquid two-phase flow. *Chemical Engineering Science*, 59(8):1853–1861.

Denev, J., Fröhlich, J., Bockhorn, H., Schwertfirm, F., and Manhart, M. 2008, Dns and les of scalar transport in a turbulent plane channel flow at low Reynolds number. In Lirkov, I., Margenov, S., and Wasniewski, J., editors, *Large-Scale Scientific Computing*, volume 4818 of *Lecture Notes in Computer Science*, 251–258. Springer Berlin Heidelberg.

Dijkhuizen, W., Roghair, I., Annaland, M. V. S., and Kuipers, J., 2010. DNS of gas bubbles behaviour using an improved 3d front tracking modelâmodel development. *Chemical Engineering Science*, 65(4):1427 – 1437.

Doychev, T. and Uhlmann, M., 2013. Settling of finite-size particles in an ambient fluid: A Numerical Study. In *8th International Conference on Multiphase Flow, ICMF, Jeju, South Korea, 2013*.

Enright, D., Fedkiw, R., Ferziger, J., and Mitchell, I., 2002. A hybrid particle level set method for improved interface capturing. *Journal of Computational Physics*, 183(1):83 – 116.

Ern, P., Risso, F., Fabre, D., and Magnaudet, J., 2012. Wake-induced oscillatory paths of bodies freely rising or falling in fluids. *Annual Review of Fluid Mechanics*, 44:97–121.

Esmaeeli, A. and Tryggvason, G., 2005. A direct numerical simulation study of the buoyant rise of bubbles at $\mathcal{O}(100)$ Reynolds number. *Physics of Fluids*, 17:093303.

Fornberg, B., 1980. A numerical study of steady viscous flow past a circular cylinder. *Journal of Fluid Mechanics*, 98:819–855.

Fröhlich, J., 2006. *Large eddy simulation turbulenter Strömungen*. Springer.

Fujiwara, A., Minato, D., and Hishida, K., 2004. Effect of bubble diameter on modification of turbulence in an upward pipe flow. *International Journal of Heat and Fluid Flow*, 25: 481 – 488.

Garcia-Villalba, M. A., Kidanemariam, G., and Uhlmann, M., 2012. DNS of vertical plane channel flow with finite-size particles: Voronoi analysis, acceleration statistics and particle-conditioned averaging. *International Journal of Multiphase Flow*, 46:54 – 74.

Garnier, C., Lance, M., and Marie, J., 2002. Mesurement of local flow characteristics in buoyancy-driven bubbly flow at high volume fraction. *Experimental Thermal and Fluid Science*, 26:811–815.

Germano, M., Piomelli, U., Moin, P., and Cabot, W. H., 1991. A dynamic subgrid-scale eddy viscosity model. *Physics of Fluids A: Fluid Dynamics (1989-1993)*, 3(7):1760–1765.

Gore, R. and Crowe, C., 1989. Effect of particle size on modulating turbulent intensity. *International Journal of Multiphase Flow*, 15(2):279 – 285.

Göz, M. F. and Sommerfeld, M. 2004, Analysis of bubble interactions in bidisperse bubble swarms by direct numerical simulation. In Sommerfeld, M., editor, *Bubbly Flows*, Heat and Mass Transfer, 175–190. Springer Berlin Heidelberg.

Griffith, B. E. and Peskin, C. S., 2005. On the order of accuracy of the immersed boundary method: Higher order convergence rates for sufficiently smooth problems. *Journal of Computational Physics*, 208(1):75–105.

Guha, A., 2008. Transport and deposition of particles in turbulent and laminar flow. *Annual Review of Fluid Mechanics*, 40:311–341.

Hagiwara, Y., 2011. Effects of bubbles, droplets or particles on heat transfer in turbulent channel flows. *Flow, Turbulence and Combustion*, 86(3-4):343–367.

Heitkam, S., Drenckhan, W., and Fröhlich, J., 2012. Packing spheres tightly: Influence of mechanical stability on close-packed sphere structures. *Physical Review Letters*, 108: 148302.

Hieber, S. E. and Koumoutsakos, P., 2005. A Lagrangian particle level set method. *Journal of Computational Physics*, 210(1):342 – 367.

Hirt, C. and Nichols, B., 1981. Volume of fluid (VOF) method for the dynamics of free boundaries. *Journal of Computational Physics*, 39(1):201 – 225.

Hohenberg, P. C. and Halperin, B. I., 1977. Theory of dynamic critical phenomena. *Rev. Mod. Phys.*, 49:435–479.

Horowitz, M. and Williamson, C., 2010. The effect of Reynolds number on the dynamics and wakes of freely rising and falling spheres. *Journal of Fluid Mechanics*, 651:251–294.

Hosokawa, S. and Tomiyama, A., 2004. Turbulence modification in gas-liquid and solid-liquid dispersed two-phase pipe flows. *International Journal of Heat and Fluid Flow*, 25 (3):489 – 498.

Hosokawa, S., Suzuki, T., and Tomiyama, A., 2012. Turbulence kinetic energy budget in bubbly flows in a vertical duct. *Experiments in Fluids*, 52(3):719–728.

Hoyas, S. and Jimenez, J., 2008. Reynolds number effects on the Reynolds-stress budgets in turbulent channels. *Physics of Fluids*, 20(10):101511.

Ilic, M., 2006. *Statistical Analysis of Liquid Phase Turbulence Based on Direct Numerical Simulations of Bubbly Flows*. PhD thesis, Fakultät für Maschinenbau, Universität Karlsruhe (TH), Report FZKA 7199, Forschungszentrum Karlsruhe.

Ilic, M., Wörner, M., and Cacuci, D. G., 2004. Balance of liquid-phase turbulence kinetic energy equation for bubble-train flow. *Journal of Nuclear Science and Technology*, 41(3): 331–338.

Ishii, M. and Zuber, N., 1979. Drag coefficient and relative velocity in bubbly, droplet or particulate flows. *AIChE Journal*, 25(5):843–855.

Jenny, M., Dusek, J., and Bouchet, G., 2004. Instabilities and transition of a sphere falling or ascending freely in a newtonian fluid. *Journal of Fluid Mechanics*, 508:201–239.

Johnson, T. A. and Pater, V. C., 1999. Flow past a sphere up to a Reynolds number of 300. *Journal of Fluid Mechanics*, 378:19–70.

Kashinsky, O., Lobanov, P., and Randin, V., 2008. The influence of a small gas addition to the structure of gas-liquid downward flow in a tube. *Journal of Engineering Thermophysics*, 17(2):120–125.

Kataoka, I., 1986. Local instant formulation of two-phase flow. *International Journal of Multiphase Flow*, 12(5):745 – 758.

Kataoka, I. and Serizawa, A., 1989. Basic equations of turbulence in gas-liquid two-phase flow. *International Journal of Multiphase Flow*, 15(5):843 – 855.

Kataoka, I., Yoshida, K., Naitoh, M., Okada, H., and Morii, T. 2012, Transport of interfacial area concentration in two-phase flow. In Amir Mesquita, I., editor, *Nuclear Reactors*.

Kempe, T. and Fröhlich, J., 2012a. An improved immersed bouldary method with direct forcing for the simulation of particle laden flow. *Journal of Computational Physics*, 231: 3663–3684.

Kempe, T. and Fröhlich, J., 2012b. Collision modelling for the interface-resolved simulation of spherical particles in viscous fluids. *Journal of Fluid Mechanics*, 709:445–489.

Kempe, T., Vowinckel, B., and Fröhlich, J., 2014. On the relevance of collision modeling for interface-resolving simulations of sediment transport in open channel flow. *International Journal of Multiphase Flow*, 58:214–235.

Kidanemariam, A. G., Chan-Braun, C., Doychev, T., and Uhlmann, M., 2013. Direct numerical simulation of horizontal open channel flow with finite-size, heavy particles at low solid volume fraction. *New Journal of Physics*, 15:025031.

Kim, I., Elghobashi, S., and Sirignano, W. A., 1993. Three-dimensional flow over two spheres placed side by side. *Journal of Fluid Mechanics*, 246:465–488.

Kim, J. and Moin, P. 1989, Transport of passive scalars in a turbulent channel flow. In Andre, J.-C., Cousteix, J., Durst, F., Launder, B., Schmidt, F., and Whitelaw, J., editors, *Turbulent Shear Flows 6*, 85–96. Springer Berlin Heidelberg.

Kim, J., Moin, P., and Moser, R., 1987. Turbulence statistics in fully developed channel flow at low Reynolds number. *Journal of Fluid Mechanics*, 177:133–166.

Krepper, E., Lucas, D., and Prasser, H.-M., 2005. On the modelling of bubbly flow in vertical pipes. *Nuclear Engineering and Design*, 235(5):597 – 611.

Krepper, E., Lucas, D., Frank, T., Prasser, H.-M., and Zwart, P. J., 2008. The inhomogeneous MUSIG model for the simulation of polydispersed flows. *Nuclear Engineering and Design*, 238(7):1690–1702.

Kurose, R. and Komori, S., 1999. Drag and lift forces on a rotating sphere in a linear shear flow. *Journal of Fluid Mechanics*, 384:183–206.

Lance, M. and Bataille, J., 1991. Turbulence in the liquid phase of a uniform bubbly air-water flow. *Journal of Fluid Mechanics*, 222:95–118.

Lange, R.-J., 2012. Potential theory, path integrals and the Laplacian of the indicator. *Journal of High Energy Physics*, 11:32.

Lelouvetel, J., Tanaka, T., Sato, Y., and Hishida, K., 2014. Transport mechanisms of the turbulent energy cascade in upward/downward bubbly flows. *Journal of Fluid Mechanics*, 741:514–542.

Lelouvetel, J., Nakagawa, M., Sato, Y., and Hishida, K., 2011. Effect of bubbles on turbulent kinetic energy transport in downward flow measured by time-resolved ptv. *Experiments in Fluids*, 50(4):813–823.

Li, W., Yan, Y., and Smith, J., 2003. A numerical study of the interfacial transport characteristics outside spheroidal bubbles and solids. *International Journal of Multiphase Flow*, 29:435–460.

Liebster, H., 1927. Über den Widerstand von Kugeln. *Annalen der Physik*, 82:541–562 (in German).

Lopez de Bertodano, M., Lahey Jr, R., and Jones, O., 1994. Phase distribution in bubbly two-phase flow in vertical ducts. *International Journal of Multiphase Flow*, 20(5):805 – 818.

Lu, J. and Tryggvason, G., 2008. Effect of bubble deformability in turbulent bubbly upflow in a vertical channel. *Physics of Fluids*, 20:040701.

Lu, J., Biswas, S., and Tryggvason, G., 2006. A dns study of laminar bubbly flows in a vertical channel. *International Journal of Multiphase Flow*, 32:643–660.

Lu, J. and Tryggvason, G., 2007. Effect of bubble size in turbulent bubbly downflow in a vertical channel. *Chemical Engineering Science*, 62(11):3008–3018.

Lu, J. and Tryggvason, G., 2013. Dynamics of nearly spherical bubbles in a turbulent channel upflow. *Journal of Fluid Mechanics*, 732:166–189.

Martinez Mercado, J., Chehata Gomez, D., Van Gils, D., Sun, C., and Lohse, D., 2010. On bubble clustering and energy spectra in pseudo-turbulence. *Journal of Fluid Mechanics*, 650:287–306.

Mendez-Diaz, S., Zenit, R., Chiva, S., Munoz-Cobo, J., and Martinez-Martinez, S., 2012. A criterion for the transition from wall to core peak gas volume fraction distributions in bubbly flows. *International Journal of Multiphase Flow*, 43:56 – 61.

Mendez-Diaz, S., Serrano-Garcia, J. C., Zenit, R., and Hernandez-Cordero, J. A., 2013. Power spectral distributions of pseudo-turbulent bubbly flows. *Physics of Fluids*, 25(4): 043303.

Menter, F. R., 1994. Two-equation eddy-viscosity turbulence models for engineering applications. *AIAA Journal*, 32(8):1598–1605.

Mercado, J. M., Prakash, V. N., Tagawa, Y., Sun, C., and Lohse, D., 2012. Lagrangian statistics of light particles in turbulence. *Physics of Fluids*, 24:055106.

Michaelides, E., 2006. *Particles, Bubbles & Drops: Their Motion, Heat and Mass Transfer.* World Scientific Publ.

Mittal, R. and Iaccarino, G., 2005. Immersed boundary methods. *Annual Review of Fluid Mechanics*, 37:239–261.

Mohd-Yusof, J., 1997. Combined immersed-boundary/B-spline methods for simulations of flow in complex geometries. *Annual Research Briefs. NASA Ames Research Center, Stanford University Center of Turbulence Research: Stanford*, 317–327.

Monchaux, R., Bourgoin, M., and Cartellier, A., 2012. Analyzing preferential concentration and clustering of inertial particles in turbulence. *International Journal of Multiphase Flow*, 40:1–18.

Nowbahar, A., Sardina, G., Picano, F., and Brandt, L., 2013. Turbophoresis attenuation in a turbulent channel flow with polymer additives. *Journal of Fluid Mechanics*, 732:706–719.

Osher, S. and Sethian, J. A., 1988. Fronts propagating with curvature-dependent speed: algorithms based on Hamilton-Jacobi formulations. *Journal of Computational Physics*, 79 (1):12–49.

Pan, D., 2010. A simple and accurate ghost cell method for the computation of incompressible flows over immersed bodies with heat transfer. *Numerical Heat Transfer, Part B: Fundamentals*, 58(1):17–39.

Pan, D., 2012. A General Boundary Condition Treatment in Immersed Boundary Methods for Incompressible Navier-Stokes Equations with Heat Transfer. *Numerical Heat Transfer, Part B: Fundamentals*, 61(4):279–297.

Panidis, T. and Papailiou, D. D., 2000. The structure of two-phase grid turbulence in a rectangular channel: an experimental study. *International Journal of Multiphase Flow*, 26 (8):1369 – 1400.

Peskin, C. S., 1977. Numerical analysis of blood flow in the heart. *Journal of Computational Physics*, 25(3):220–252.

Politano, M., Carrica, P., and Converti, J., 2003. A model for turbulent polydisperse two-phase flow in vertical channels. *International Journal of Multiphase Flow*, 29(7):1153 – 1182.

Pope, S. B., 2000. *Turbulent Flows*. Cambridge University Press.

Prandtl, L. and Wieghardt, K., 1947. *Über ein neues Formelsystem für die ausgebildete Turbulenz*. Vandenhoeck & Ruprecht.

Roghair, I., Lau, Y., Deen, N., Slagter, H., Baltussen, M., Annaland, M. V. S., and Kuipers, J., 2011. On the drag force of bubbles in bubble swarms at intermediate and high Reynolds numbers. *Chemical Engineering Science*, 66(14):3204 – 3211.

Roghair, I., Baltussen, M., Van Sint Annaland, M., and Kuipers, J., 2013. Direct numerical simulations of the drag force of bi-disperse bubble swarms. *Chemical Engineering Science*, 95:48–53.

Roma, A. M., Peskin, C. S., and Berger, M. J., 1999. An adaptive version of the immersed boundary method. *Journal of computational physics*, 153(2):509–534.

Roussel, J., 2014. Turbulent Kinetic Energy Budget for Bubbly Flows. Master's thesis, Technische Universität Dresden, Germany and Ecole Polytechnique, France.

Sagaut, P., 2006. *Large Eddy Simulation for Incompressible Flows: An Introduction*. Springer.

Santarelli, C. and Fröhlich, J., 2014. Direct Numerical Simulation of spherical bubbles in vertical turbulent channel flow. *International Journal of Multiphase Flow (under review)*.

Santarelli, C., Kazutaka, H., Fukagata, K., and Fröhlich, J., 2014a. A smoothed void fraction approach for the detection of bubble clusters. In *Proceedings of the ASME 2014 4th Joint US-European Fluids Engineering Division Summer Meeting*, number Paper No. FEDSM2014-21244.

Santarelli, C., Kempe, T., and Fröhlich, J., 2014b. An Immersed Boundary Method and a Ghost Cell Method for the Simulation of Heat Transfer Problems. In B. Sarler, N. M. and Nithiarasu, P., editors, *Proceedings of the 3th International Conference on Computational Methods for Thermal Problems*.

Santarelli, C. and Fröhlich, J., 2013. Characterisation of bubble clusters in simulations of upward turbulent channel flow. *Proceedings in Applied Mathematics and Mechanics*, 13 (1):19–22.

Sato, Y., Sadatomi, M., and Sekoguchi, K., 1981. Momentum and heat transfer in two-phase bubble flow - I. theory. *International Journal of Multiphase Flow*, 7(2):167–177.

Schwarz, S., 2014. *An immersed boundary method for particles and bubbles in magnetohydrodynamic flows*. PhD thesis, Technical University Dresden.

Schwarz, S. and Fröhlich, J., 2012. Representation of deformable bubbles by analytically defined shapes in an immersed boundary method. *AIP Conference Proceedings*, 1479(1): 104–108.

Schwarz, S. and Fröhlich, J., 2014. Numerical study of single bubble motion in liquid metal exposed to a longitudinal magnetic field. *International Journal of Multiphase Flow*, 62: 134–151.

Schwarz, S., Kempe, T., and Fröhlich, J., 2015. A temporal discretization scheme to compute the motion of light particles in viscous flows by an immersed boundary method. *Journal of Computational Physics*, 281:591–613.

Seo, J. H. and Mittal, R., 2011. A high-order immersed boundary method for acoustic wave scattering and low-mach number flow-induced sound in complex geometries. *Journal of computational physics*, 230(4):1000–1019.

Serizawa, A., Kataoka, I., and Michiyoshi, I., 1975a. Turbulence structure of air-water bubbly flow -II. Local properties. *International Journal of Multiphase Flow*, 2:235 – 246.

Serizawa, A., Kataoka, I., and Michiyoshi, I., 1975b. Turbulence structure of air-water bubbly flow - III. Transport properties. *International Journal of Multiphase Flow*, 2:247–259.

Shawkat, M. E. and Ching, C., 2011. Liquid turbulence kinetic energy budget of co-current bubbly flow in a large diameter vertical pipe. *Journal of Fluids Engineering*, 133(9): 091303.

Sint Annaland, M. V., Dijkhuizen, W., Deen, N., and Kuipers, J., 2006. Numerical simulation of behavior of gas bubbles using a 3-D front-tracking method. *AIChE Journal*, 52(1):99–110.

Smagorinsky, J., 1964. General circulation experiments with the primitive equations. *Monthly Weather Review*, 91:99–164.

Spalart, P. R. and Allmaras, S. R., 1994. A one-equation turbulence model for aerodynamic flows. *Recherche Aerospatiale*, 1:5–21.

Sun, X., Paranjape, S., Kim, S., Goda, H., Ishii, M., and Kelly, J. M., 2004. Local liquid velocity in vertical air-water downward flow. *Journal of Fluids Engineering*, 126(4):539–545.

Sussman, M. and Puckett, E. G., 2000. A coupled level set and volume-of-fluid method for computing 3d and axisymmetric incompressible two-phase flows. *Journal of Computational Physics*, 162(2):301 – 337.

Sussman, M., Smereka, P., and Osher, S., 1994. A level set approach for computing solutions to incompressible two-phase flow. *Journal of Computational Physics*, 114(1):146–159.

Tagawa, Y., Ogasawara, T., Takagi, S., and Matsumoto, Y., 2010. Surfactant effects on single bubble motion and bubbly flow structure. *AIP Conference Proceedings*, 1207(1): 43–48.

Takagi, S., Ogasawara, T., and Y., M., 2008. The effects of surfactant on the multiscale structure of bubbly flows. *Philosophical Transactions of the Royal Society A*, 366:2117–2129.

Takemura, F. and Magnaudet, J., 2003. The transverse force on clean and contaminated bubbles rising near a vertical wall at moderate Reynolds number. *Journal of Fluid Mechanics*, 495:235–253.

Tanaka, M. 2011, Numerical study on flow structures and heat transfer characteristics of turbulent bubbly upflow in a vertical channel. In Jianping Zhu, I., editor, *Computational Simulations and Applications*, 119–142.

Tanaka, M., Matsui, N., Miyajima, Y., and Hagiwara, Y., 2010. Characteristics of heat transfer in turbulent bubbly upflow in a vertical channel. In *7th International Conference on Multiphase Flow, ICMF 2010*.

Tanaka, T. and Eaton, J. K., 2010. Sub-kolmogorov resolution partical image velocimetry measurements of particle-laden forced turbulence. *Journal of Fluid Mechanics*, 643:177–206.

Tasoglu, S., Demirci, U., and Muradoglu, M., 2008. The effect of soluble surfactant on the transient motion of a buoyancy-driven bubble. *Physics of Fluids*, 20(4):040805.

Tavassoli, H., Kriebitzsch, S., van der Hoef, M., Peters, E., and Kuipers, J., 2013. Direct numerical simulation of particulate flow with heat transfer. *International Journal of Multiphase Flow*, 57:29 – 37.

Titscher, T., 2012. Vergleich von Immersed Boundary Methoden für die Temperatur. Master's thesis, Technische Universität Dresden, Institute of Fluid Mechanics.

Tomiyama, A., Tamai, H., Zun, I., and Hosokawa, S., 2002. Transverse migration of single bubbles in simple shear flows. *Chemical Engineering Science*, 57(11):1849–1858.

Tran-Cong, S., Marie, J., and Perkins, R., 2008. Bubble migration in a turbulent boundary layer. *International Journal of Multiphase Flow*, 34(8):786 – 807.

Troshko, A. and Hassan, Y., 2001. A two-equation turbulence model of turbulent bubbly flows. *International Journal of Multiphase Flow*, 27(11):1965 – 2000.

Uhlmann, M., 2005. An immersed boundary method with direct forcing for the simulation of particulate flows. *Journal of Computational Physics*, 209:448–476.

Uhlmann, M., 2008. Interface-resolved Direct Numerical Simulation of vertical particulate channel flow in the turbulent regime. *Physics of Fluids*, 20:053305.

Uhlmann, M., 2007. Investigating turbulent particulate channel flow with interface-resolved DNS. In *6th International Conference on Multiphase Flow, ICMF, Leipzig, Germany 2007*.

Unverdi, S. and Tryggvason, G., 1992. A front-tracking method for viscous, incompressible, multi-fluid flows. *Journal of Computational Physics*, 100(1):25–37.

Van Sint Annaland, M., Deen, N., and Kuipers, J., 2005. Numerical simulation of gas bubbles behaviour using a three-dimensional volume of fluid method. *Chemical Engineering Science*, 60(11):2999–3011.

Wang, S., Lee, S., Jones, O., and Lahey, R., 1987. 3-D turbulence structure and phase distribution measurements in bubbly two-phase flows. *International Journal of Multiphase Flow*, 13(3):327 – 343.

Wang, Z., Fan, J., Luo, K., and Cen, K., 2009. Immersed boundary method for the simulation of flows with heat transfer. *International Journal of Heat and Mass Transfer*, 52(19-20): 4510 – 4518.

Wieselsberger, C., 1921. Neuere Feststellungen über die Gesetze des Flüssigkeits- und Luftwiderstandes. *Physikalische Zeitschrift*, 23:321–328 (in German).

Xia, J., Luo, K., and Fan, J., 2014. A ghost-cell based high-order immersed boundary method for inter-phase heat transfer simulation. *International Journal of Heat and Mass Transfer*, 75:302 – 312.

Yamamoto, Y. and Kunugi, T., 2011. Direct Numerical Simulation of Turbulent Channel Flow with Deformed Bubbles. *Progress in Nuclear Science and Technology*, 2:543–549.

Young, J. and Leeming, A., 1997. A theory of particle deposition in turbulent pipe flow. *Journal of Fluid Mechanics*, 340:129–159.

Youngs, D., 1982. Time-dependent multi-material flow with large fluid distortion. *Numerical methods for fluid dynamics*, 24:273–285.

Yue, P., Feng, J. J., Liu, C., and Shen, J., 2004. A diffuse-interface method for simulating two-phase flows of complex fluids. *Journal of Fluid Mechanics*, 515:293–317.

Zeng, L., Balachendar, S., Fischer, P., and Najjar, F., 2008. Interactions of a stationary finite-sized particle with wall turbulence. *Journal of Fluid Mechanics*, 594:271–305.

Zhang, N., Zheng, Z., and Eckels, S., 2008. Study of heat-transfer on the surface of a circular cylinder in flow using an Immersed-Boundary Method. *International Journal of Heat and Fluid Flow*, 29(6):1558–1566.

A   Correction of the pressure field in the vicinity of the phase boundary

One of the feature of the interpolation scheme with regularized delta function employed in the present work is that fluid quantities in the proximity of the phase boundary are smoothed and do not present a sharp gradient over the phase boundary, as in other numerical approaches such as the VOF. This feature is of great advantages when simulating moving objects, as reported by Uhlmann (2005) and Kempe and Fröhlich (2012a), since time oscillations of forces are avoided due to the smoothed properties of the interpolation scheme. Nevertheless, this approach presents also some drawbacks: Fluid properties such as velocity and pressure on the Eulerian mesh are not exact in the proximity of the phase boundary, but are only correct in a "smoothed" sense. The region where fluid properties present such values is limited to a region between $r_p + 1.5\Delta$ and $r_p - 1.5\Delta$, since the δ-functions employ a stencil which extends over three Eulerian points. The pressure field is more affected by this problem than the velocity field, since the forcing of the IBM is calibrated on the desired velocity to be imposed on the phase boundary, while the pressure field is modified by the correction step in the NSE. This feature implies that the evaluation of fluid forces on the phase boundary does not yield correct values if points in the vicinity of the phase boundary are employed. Several methods have been proposed to overcome this problems. Schwarz and Fröhlich (2012) and Schwarz (2014) developed an algorithm to simulate deformable bubbles where the deformation is a function of the pressure at the phase boundary. These authors proposed to evaluate the pressure not on the phase boundary, but on a surface that is shifted in the radial direction (in the fluid region) of a distance equal 1.5Δ to exclude the influence of the Eulerian points inside the bubbles in the evaluation of the pressure on the phase boundary. This approach, although providing a better estimation of the pressure, slightly underestimate the pressure level on the phase boundary. Therefore, in the present work, the pressure on the Eulerian points in the proximity of the phase boundary is evaluated in a different manner.

The proposed method involves only Eulerian points in the proximity of the phase boundary, i.e. for which the radial distance from the phase boundary is below $\sqrt{2}\Delta$. These points are now defined as Ghost Points, GP, as in the Ghost Cell Method (GCM) described in Sec. 8.2.2. The difference with respect to the CGM is that now the GP are both inside and outside the phase boundary, as portrayed in Fig. A.1. Two sets of Image Points are defined: One set, labeled IP1, at distance δ from the phase boundary and the second set, labeled IP2, at distance 2δ from the phase boundary. The distance δ was set equal to the mesh step size Δ. The pressure at IP1 and IP2 is obtained by linear interpolation from the eight (four) surrounding points in the three (two) dimensional case. Once the pressure at IP1 and IP2 is evaluated, it can be linearly extrapolated to the GP, as depicted in Fig. A.2.

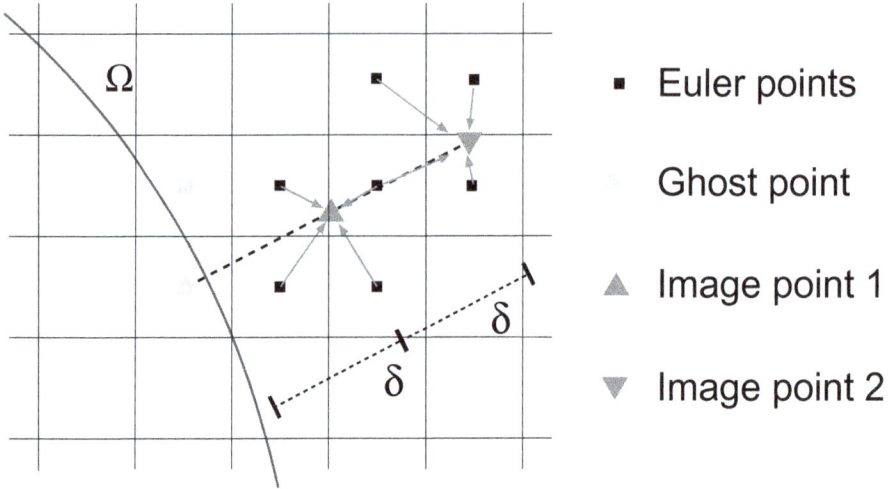

Figure A.1: Points involved in the evaluation of the pressure in the proximity of the phase boundary.

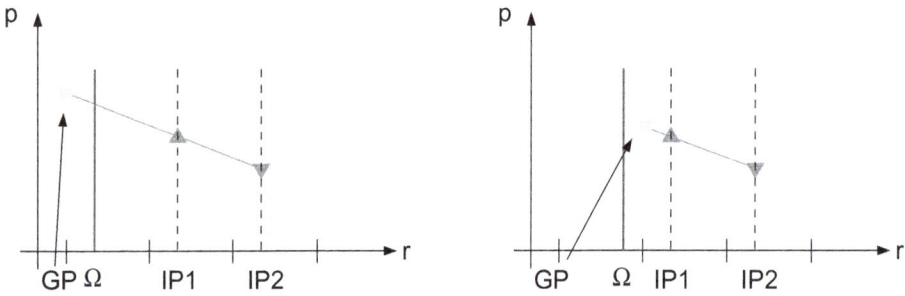

Figure A.2: Linear extrapolation of the pressure from IP1 and IP2 to GP: Schematic representation in radial direction. Symbols as in Fig. A.1. Left: GP inside the phase boundary. Right: GP ouside the phase boundary.

This approach is validated for a two-dimensional flow around a cylinder and for a three-dimensional flow around a sphere. Both validations are now briefly described.

The first simulation approaches the flow around a cylinder at $Re_{p,c} = 40$, based on the diameter d_p and on the inflow velocity u_c. Figure A.3 portrays, for $y/L < 0.5$, the unmodified pressure field in the proximity of the cylinder surface and, for $y/L > 0.5$, the corrected pressure field. After the pressure values on the GP have been overwritten, the pressure at the discrete points of the phase boundary was interpolated by means on bilinear interpolation and then compared with the results of Fornberg (1980), as depicted in Fig. A.4, by means of the pressure coefficient defined as

$$C_p = \frac{p - p_{ref}}{1/2 \, u_c^2 \, \rho} \, . \tag{A.1}$$

A slight underestimation of the pressure at the front stagnation point is observed, most probably due to the approximation of the pressure field as a linear function in radial direction.

Nevertheless, a very good agreement is found between the results provided by the present approach and the data in the literature.

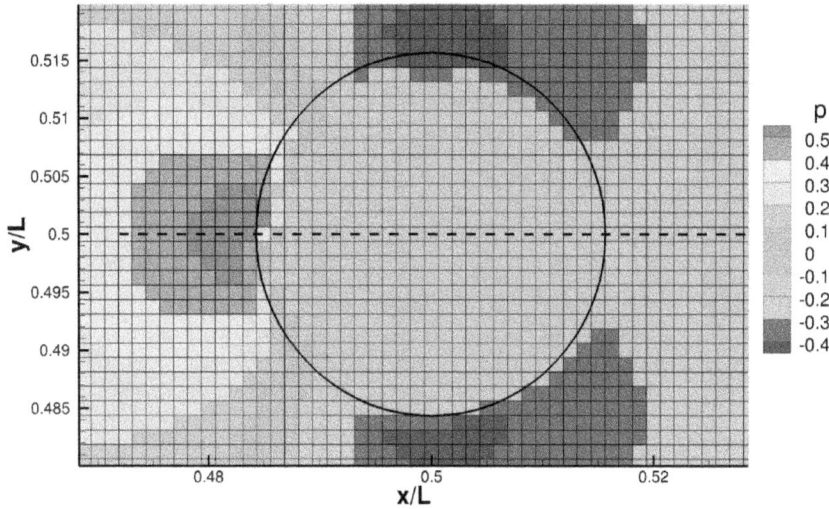

Figure A.3: Pressure field in the proximity of the cylinder surfaces. For $y/L < 0.5$, no correction of the pressure is employed. For $y/L > 0.5$, the pressure is corrected and the horizontal dashed line separates the two regions.

The second problem investigated is the flow around a fixed sphere at $Re_{p,c} = 100$ based on the diameter and on the inflow velocity. For this value of the Reynolds number, the wake is axisymmetric with respect to the streamwise direction. The distribution of C_p on the phase boundary according to (A.1) is compared with the results of Li et al. (2003). As in the previous case, the agreement with results provided by a simulation with body-fitted mesh is very good, even if the pressure coefficient is slightly underestimated at the front stagnation point, as portrayed in Fig. A.5.

In conclusion, a method was proposed and validated to evaluate the pressure field in the proximity of the phase boundary. This method does not modify the pressure field during the simulation, but is only employed in the post-processing phase and therefore does not imply any modifications of the IBM itself.

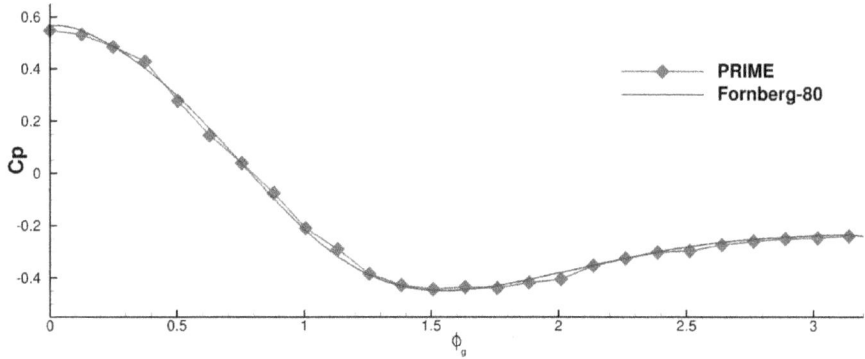

Figure A.4: Pressure coefficient at the phase boundary of a fixed cylinder: Data from (Fornberg, 1980) and modified pressure field by PRIME. The angle ϕ_g is defined starting from the front stagnation points. The ratio between sphere diameter and mesh step size is 30.

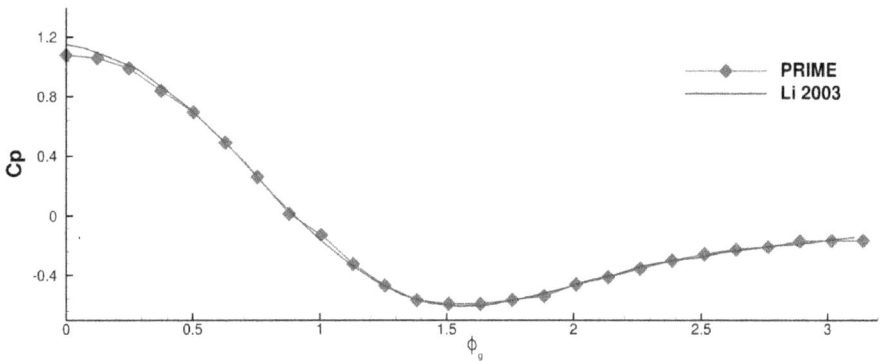

Figure A.5: Pressure coefficient at the phase boundary of a fixed sphere. The angle ϕ_g is defined starting from the front stagnation points. The ratio between sphere diameter and mesh step size is 25.

B Budget of local instant turbulent kinetic energy

This section reports on the evaluation and visualization of the local instantaneous turbulent kinetic energy for a single-phase flow. An average field $\langle u \rangle$ needs to be defined, where the average operator depends on specific problem. For a channel flow, usually $\langle u \rangle = \langle u \rangle_{xzt} = \langle u \rangle(y)$ The fluctuating field is hence defined as

$$u'(x, y, z, t) = u(x, y, z, t) - \langle u \rangle(y) \tag{B.1}$$

From the NSE of u_i and from the transport equation of the average field $\langle u_i \rangle$, the transport equation of u_i' is found

$$\frac{\partial u_i'}{\partial t} + u_j \frac{\partial u_i'}{\partial x_j} = -\frac{\partial}{\partial x_j} \langle u_i' u_j' \rangle - u_j' \frac{\partial \langle u_i \rangle}{\partial x_j} + \frac{\partial \tau_{ij}'}{\partial x_j} - \frac{1}{\rho} \frac{\partial p'}{\partial x_i} \tag{B.2}$$

where

$$\tau_{ij}' = \nu \left(\frac{\partial u_i'}{\partial x_j} + \frac{\partial u_j'}{\partial x_i} \right). \tag{B.3}$$

Multiplying (B.2) for u_i' yields

$$u_i' \frac{\partial u_i'}{\partial t} + u_i' u_j \frac{\partial u_i'}{\partial x_j} = -u_i' \frac{\partial}{\partial x_j} \langle u_i' u_j' \rangle - u_i' u_j' \frac{\partial \langle u_i \rangle}{\partial x_j} + u_i' \frac{\partial \tau_{ij}'}{\partial x_j} - \frac{1}{\rho} u_i' \frac{\partial p'}{\partial x_i} \tag{B.4}$$

with

$$\tilde{K} = \frac{1}{2}(u_i' u_i'). \tag{B.5}$$

The quantity $\tilde{K}$ is defined as the *local instantaneous* turbulent kinetic energy. The first term of LHS (LSH1) becomes

$$u_i' \frac{\partial u_i'}{\partial t} = \frac{\partial u_i' u_i'}{\partial t} - u_i' \frac{\partial u_i'}{\partial t} = \frac{\partial \tilde{K}}{\partial t} \tag{B.6}$$

and, similarly, (B.4) becomes

$$\frac{\partial \tilde{K}}{\partial t} + u_j \frac{\partial \tilde{K}}{\partial x_j} = -\frac{\partial}{\partial x_j} \left(u_i' \langle u_i' u_j' \rangle \right) + \langle u_i' u_j' \rangle \frac{\partial u_i'}{\partial x_j} - u_i' u_j' \frac{\partial \langle u_i \rangle}{\partial x_j} + $$
$$\frac{\partial}{\partial x_j} \left(u_i' \tau_{ij}' \right) - \tau_{ij}' \frac{\partial u_i'}{\partial x_j} - \frac{\partial}{\partial x_i} \left(p' u_i' \right) + p' \frac{\partial u_i'}{\partial x_i} \tag{B.7}$$

where the last term (RHS7) is zero due to continuity of fluctuating field. The TKE is commonly defined as

$$K = \left\langle \tilde{K} \right\rangle = \frac{1}{2} \langle u_i' u_i' \rangle \tag{B.8}$$

and, in channel flow configuration, usually $K = K(y)$. It has to be noted that some terms contain subterms that are equal to zero

$$\frac{\partial \langle u_i \rangle}{\partial x} = \frac{\partial \langle u_i \rangle}{\partial z} = \frac{\partial \langle u_i \rangle}{\partial t} = 0 \tag{B.9}$$

Averaging (B.7) yields the well-know transport equation of K, where RHS2 of (B.7) goes to zero due to the definition of the fluctuating field.

The transport equation of K reads

$$\frac{\partial K}{\partial t} + \langle u_j \rangle \frac{\partial K}{\partial x_j} = -\frac{\partial}{\partial x_j} \langle u_i' u_i' u_j' \rangle - \langle u_i' u_j' \rangle \frac{\partial \langle u_i \rangle}{\partial x_j} + \\ \frac{\partial}{\partial x_j} \langle \tau_{ij}\, u_i' \rangle - \left\langle \tau_{ij} \frac{\partial u_i'}{\partial x_j} \right\rangle - \frac{\partial}{\partial x_i} \langle p' u_i' \rangle \tag{B.10}$$

where RHS1 + RHS3 + RHS5 are jointly ofter referred to as "transport" or "diffusion term" C, RHS2 as the product term Π and RHS4 as the dissipation term ϵ, and due to average, each term is a function of y only. For stationary steady state

$$\frac{\partial K}{\partial t} + \langle u_j \rangle \frac{\partial K}{\partial x_j} = 0 \tag{B.11}$$

and

$$\Pi + \epsilon + C = 0 \,. \tag{B.12}$$

Looking at each term of (B.7) we may define an instantaneous production as

$$\tilde{\Pi}(x,y,k,t) = -u_i' u_j' \frac{\partial \langle u_i \rangle}{\partial x_j} = -u'(x,y,k,t)\, v'(x,y,k,t) \frac{\partial \langle u \rangle (y)}{\partial y} \,. \tag{B.13}$$

For the dissipation term,

$$\tilde{\epsilon}(x,y,k,t) = \tau_{ij}' \frac{\partial u_i'}{\partial x_j} \tag{B.14}$$

For the transport term,

$$\tilde{C}(x,y,k,t) = \frac{\partial}{\partial x_j} \left[-u_i' \langle u_i' u_j' \rangle + u_i' \tau_{ij}' - (u_i' p')\delta_{ij} \right] + \langle u_i' u_j' \rangle \frac{\partial u_i'}{\partial x_j} \tag{B.15}$$

so that (B.7) can be written as $\forall$

$$\frac{\partial \tilde{K}}{\partial t} + u_j \frac{\partial \tilde{K}}{\partial x_j} = \tilde{\Pi} - \tilde{\epsilon} + \tilde{C} \,. \tag{B.16}$$

Obviously, averaging (B.16) yields (B.10). Visualizations of the aforementioned terms are shown in Figs. B.1-B.3 below.

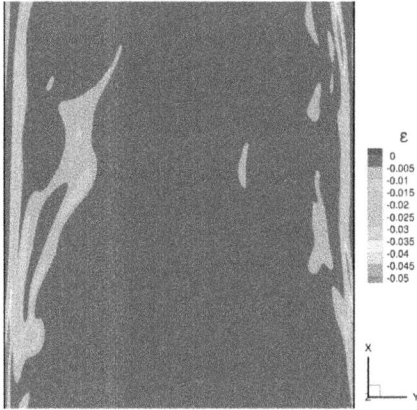

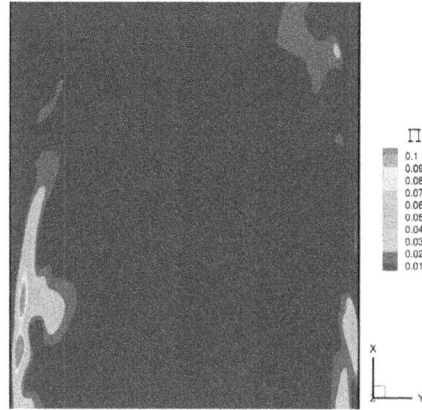

Figure B.1: Local dissipation field for case *Unladen*: Contours represented on a vertical wall at $z/H = L_z$. The whole channel width is portrayed and the walls are represented by the black vertical lines.

Figure B.2: Local production field for case *Unladen*: contours represented on a vertical wall at $z/H = L_z$. The whole channel width is portrayed and the walls are represented by the black vertical lines.

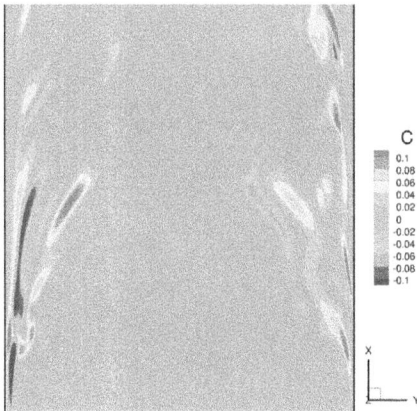

Figure B.3: Local transport term for case *Unladen*: contours represented on a vertical wall at $z/H = L_z$. The whole channel width is portrayed and the walls are represented by the black vertical lines.

www.ingramcontent.com/pod-product-compliance
Lightning Source LLC
Chambersburg PA
CBHW081532220326
41598CB00036B/6414